Regina de Amorim Romacheli

Environmental Impact Assessment: Theory and Practice

Regina de Amorim Romacheli

Environmental Impact Assessment: Theory and Practice

A practical approach to environmental studies

ScienciaScripts

Imprint

Any brand names and product names mentioned in this book are subject to trademark, brand or patent protection and are trademarks or registered trademarks of their respective holders. The use of brand names, product names, common names, trade names, product descriptions etc. even without a particular marking in this work is in no way to be construed to mean that such names may be regarded as unrestricted in respect of trademark and brand protection legislation and could thus be used by anyone.

Cover image: www.ingimage.com

This book is a translation from the original published under ISBN 978-613-9-71883-2.

Publisher:
Sciencia Scripts
is a trademark of
Dodo Books Indian Ocean Ltd. and OmniScriptum S.R.L publishing group

120 High Road, East Finchley, London, N2 9ED, United Kingdom
Str. Armeneasca 28/1, office 1, Chisinau MD-2012, Republic of Moldova, Europe
Printed at: see last page
ISBN: 978-620-7-96049-1

ACKNOWLEDGEMENTS

To God, who blesses me, keeps me and guides me;

To my parents, for all the affection and love they have shown me, for the wisdom they have always shown me and for the family they have given me;

To my supervisor, Prof Dr Jorge Madeira Nogueira, for his valuable contributions;

To my teachers at CEEMA, for the knowledge they imparted;

To my classmates, for their companionship and friendship;

To Waneska, a CEEMA employee, for her support and attention;

To my siblings, Angélica, Norma and Leandro, and my nephews, Anna Cleide, Pedro, Lucca, Lara, Lina and Liz, for the love and joy they give me.

SUMMARY

Environmental impact assessment is an environmental policy instrument that aims to identify, evaluate, predict and mitigate the biophysical, social and other relevant effects of project proposals. This dissertation studies the Environmental Impact Assessments presented in Brazil, discussing how impact assessment methods are used and their implications for the effectiveness of EIA as an environmental policy. The work encompasses an extensive literature review and an investigation into the operation of these methods, seeking to identify weaknesses and potential in relation to the elements most discussed internationally: cumulative effects and significance.

Keywords: environmental impact assessment (EIA), environmental impact assessment methods.

SUMMARY

CHAPTER 1   5

CHAPTER 2   8

CHAPTER 3   34

CHAPTER 4   48

CHAPTER 5   62

CHAPTER 6   77

CHAPTER 7   79

CHAPTER 1

INTRODUCTION

Environmental impact assessment is a command and control instrument created in Brazil by Law No. 6.938/81 establishing the National Environmental Policy. Its main objective is to identify, assess, predict and mitigate the biophysical, social and other relevant effects of proposed projects and physical activities before major decisions and commitments are made (SADLER, 1996; MOREIRA, 1992; CHRISTA, 2005).

Considering the need to establish definitions, responsibilities, basic criteria and general guidelines for the use and implementation of Environmental Impact Assessment as one of the instruments of the National Environmental Policy, CONAMA Resolution 001/86 instituted the **Environmental Impact Study and the Environmental Impact Report - EIA/RIMA.** These studies are intended to encompass in technical-scientific language all the environmental policy guidelines that will guide the government's decision on whether or not to allow the implementation of the activity, making it subject to environmental licensing.

In Brazil, the term Environmental Impact Assessment (EIA) is now used to describe a technical activity to be carried out in the EIS, which, according to the wording of article 5, item II, aims to **"systematically identify and assess the environmental impacts generated during the implementation and operation phases of the activity".**

Impacts are identified using[1] **environmental impact assessment methods,** which are designed to provide a reading of the impacts in relation to the environmental elements considered relevant (magnitude, importance, reversibility, duration, etc.), subsidising a subsequent assessment of potential impacts and the consequent description of measures and monitoring to minimise the environmental damage caused by the project. Each method described in the literature encompasses different variables and has different operationalisations.

The aim of this dissertation is to verify the hypothesis that **the way in which environmental impacts are analysed within an EIA reflects the degree of alteration caused by a project to the natural base.** To this end, the work follows a logical sequence of reasoning, which allows for a better understanding of the subject.

---

[1] Although most of the authors referenced in this dissertation use the term **environmental impact assessment methodology,** we believe that the correct term for what we intend to study is **environmental impact assessment method,** since it refers to the description of each method and not the study of methods (methodology).

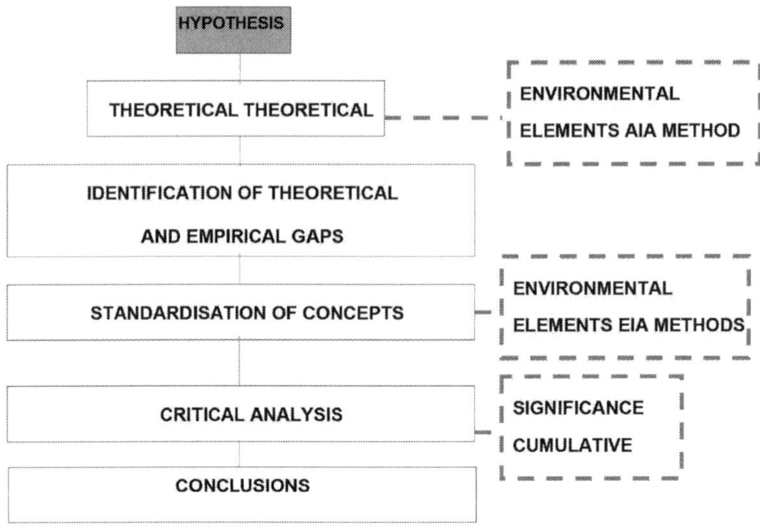

Figure 1.0 - Work development sequence

Source: Prepared by the author

The work is based on scientific research into the subject, focussing on two approaches: **theoretical and empirical research.** Theoretical research will be carried out in national and international literature, identifying the state of the art and, as a **first contribution,** identifying gaps that may hinder understanding. The focus of the study will be: 1 - the environmental elements to be considered and 2 - the methods available to carry out the evaluation.

The empirical research will be carried out on 8 (eight) EIAs submitted to the Brazilian Institute for the Environment and Renewable Natural Resources - IBAMA,

as a prerequisite for the environmental licensing of activities related to the electricity sector. The aim of this investigation is to verify how they are drawn up, the similarities in relation to the guidelines described in the literature, the methods used, the environmental elements considered, and the way in which the environmental impact is analysed. **As a contribution,** it also seeks to verify the operational and conceptual flaws in the instrument, which could lead to mistaken public decisions.

The EIAs discussed in this dissertation, as previously mentioned, are linked to the electricity sector, more specifically: one for a hydroelectric dam, two for transmission lines, one for a radioactive waste dump, one for an electricity booster, two for a gas pipeline and one for a gas treatment plant. There are five different preparation companies, each made up of a different multidisciplinary team. Even the EIAs drawn up by the same company differ in terms of team compared to the other studies. The multidisciplinary nature of the professional preparers identified in the studies adds value to the work as they provide different visions and operationalisations of the impact assessment, enriching the research.

As a **third contribution,** the conceptual gaps found in both the theoretical and empirical approaches will be standardised, aligning reasoning and formatting more precise concepts. This will allow the critical analysis to be supported by a solid base of information, making it more structured and plausible (chapters II and III).

The **main contribution of** this work is to verify the effectiveness of the environmental impact assessment methods used in Brazil in relation to two of the environmental elements considered most relevant in the literature: **significance,** because it is the synthesis of the entire analysis, and **cumulativeness,** because its lack is identified as the study's main flaw. To this end, chapters IV and V check whether the methods used in EIAs meet the guidelines of the scientific literature, which would make the hypothesis true.

The aim is to help improve the evaluation procedures for public policies and, as a result, to ensure that these policies are better designed and implemented.

CHAPTER 2

ENVIRONMENTAL IMPACT ASSESSMENT (
EIA) AND
ENVIRONMENTAL IMPACT STUDY
(EIA)

2.1 EIA as a public policy instrument

In 1969, in response to pressure from environmental groups, the US government instituted the *National Environmental Policy Act*, known as NEPA, regulating Environmental Impact Assessment (EIA) as public policy for the first time (GILPIN, 1995). The purpose of the EIA was to identify, evaluate, predict and mitigate the biophysical, social and other relevant effects of proposed projects and physical activities before major decisions and commitments were made (SADLER, 1996; MOREIRA, 1992; CHRISTA, 2005).

According to Gilpin (1995), EIA first spread to England, Germany and most of the Nordic countries, which adapted American legislation and created regulations to manage zoning and land use, including the environmental impact assessment process.

In the European Community (EU), EIA can be found in the first three environmental action programmes that the organisation developed after the 1972 United Nations Conference in Stockholm. Specifically, the principle of this policy was the creation of pollution prevention at source, considering the environmental implications of projects before their development. In 1985, the Council of the European Community approved the "EIA Directive", in which the Member States were obliged to enact regulations and apply the provisions by July 1988. The purpose of the EIA Directive was to link the "approval" of the proposed project to an authorisation issued by the environmental agency (REDEY and KISS, 1998).

According to Moreira (1992), the path of EIA in Third World countries began to be traced by international economic cooperation agents who, driven by international environmental defence groups and public opinion, began to demand that the projects they financed take environmental variables into account.

According to Christa (2005), Environmental Impact Assessment (EIA) is often described as a tool for sustainable development (SADLER 1996). For Gilpin (1995), the idea of an economic optimum, presented by Vilfredo Pareto (1848-1923), still prevails in EIA. The author considers that in a static model of resource allocation, the allocation is said to be efficient if it is not possible to rearrange this allocation and improve the well-being (the level of satisfaction) of at least one individual in society without reducing the well-being of at least one other individual (MUELLER, 2001).

The EIA is a command and control instrument that relies on direct regulation, accompanied by supervision and sanctions for non-compliance with established norms and standards (NOGUEIRA E PEREIRA, 1999). It is made up of a set of procedures capable of ensuring, right from the start of the process, that there is a systematic examination of the environmental impacts of a proposed action (project, programme, plan or policy) and its alternatives, and that the results are presented in an appropriate manner to the public and decision-makers (QUEIROZ, 1992).

According to the *International Association for Impact Assessment - IAIA* (2004), Environmental Impact Assessment must be: 1 - **intentional,** subsidising the decision-making process and resulting in adequate levels of environmental protection and social welfare; 2 - **rigorous,** applying the science of the "best possible", using appropriate methods and techniques to solve the problem being investigated[2] ; 3 - **practical,** aiming to implement viable environmental solutions, so that it can be applied by the proponents; 4 - **relevant,** providing reliable information, subsidising planning and decision-making; 5 - **cost-effective,** achieving the EIA's objectives within the limits of time, resources and methods; 6 - **efficient,** imposing burdens in terms of time and funding at minimum cost; 7 - **focused,** concentrating on significant environmental effects and fundamental issues; 8 - **easily adaptable,** adjusting to reality without compromising the integrity of the process; 9 - **participatory,** providing adequate opportunities to inform and involve stakeholders, affected public bodies and their contributors; 10 - **interdisciplinary,** ensuring that there are appropriate techniques to be employed; 11 - **credible,** carried out with professionalism, rigour, impartiality, objectivity and balance; 12 - **integrated,** addressing the interrelationships of social and economic rights and biophysical aspects; 13 - **transparent,** with content that is easy to understand, identifying controversial factors and recognising the limitations and difficulties of the proposed project and; 14 - **systematic,** considering all relevant information about the affected environment, proposed alternative locations and their impacts and the measures needed to monitor and investigate residual effects.

Internationally, the guidelines for implementing and operating EIA are similar and demonstrate a public policy that integrates the participation of the state and society in

---

[2] For Alton and Underwood (2003) the solution must be: (1) scientifically correct, (2) easy to understand, (3) possible to implement, (4) legally defensible and (5) implemented at the right time.

decision-making (SADLER, 1996; GILPIN, 1995 and IAIA, 2004). They are subdivided into four stages:

- 1º step: assessment of the need to carry out an EIA; if so, a term of reference is drawn up indicating all the studies and procedures necessary for the approval of the project;
- 2º step: carrying out the impact analysis with the determination of mitigating measures, the indication of the environmentally viable alternative and the acceptability of the proposal;
- 3º step: presentation of the EIA, containing in an impartial manner the impacts, measures and wishes of the society affected, and;
- 4º step: analysing the EIA and making decisions.

2.2 EIA in Brazil: applications and shortcomings

According to Moreira (2002), the first environmental assessment carried out in Brazil dates back to 1972, when the World Bank financed the Sobradinho dam and hydroelectric power station. According to the author, until 1986, a considerable number of projects dependent on external funding were subject to EIA, although the results of the studies were not submitted to the environmental control bodies.

The pioneering regulation for the use of EIA in Brazil dates back to 1977, at state level. The act regulating the Licensing System for Polluting Activities (SLAP) by the government of the state of Rio de Janeiro established provisions for the State Commission for Environmental Control to require, when it deemed it necessary, the preparation and presentation of an Environmental Impact Report (RIMA) to technically instruct the application for any type of licence (MOREIRA, 2002; SANCHEZ, 2006).

The regulation of Environmental Impact Assessment as a public policy instrument was established by Federal Law No. 6.938 of 31 August 1981, which provides for the National Environmental Policy. The proposed instrument seeks to meet the objectives of the policy, mainly in relation to: making economic and social development compatible, defining priority areas for government action, establishing criteria and standards for environmental quality, preserving environmental resources and maintaining ecological balance and imposing on the polluter or predator to recover/indemnify the damage (LAW No. 6.938/81, article 4).

In Brazil there is a particularity: the EIA, instituted at federal level, is linked to environmental licensing. The licensing of an activity that modifies the environment depends on the preparation and approval of the EIA and the respective RIMA, which are part of the EIA. For Sánchez (2006), such a direct relationship between EIA and licensing was a strategy employed by the technicians involved to facilitate acceptance of a new environmental planning tool and establish a context of application that was already familiar, i.e. environmental licensing. (MOREIRA, 2002; MILARÉ, 2000 *apud* NICOLAÍDES, 2005)

The Environmental Impact Assessment (EIA), according to Moreira (1992), arose from the establishment of the National Environmental Policy Act (NEPA), which stipulates that all proposals and actions of the federal government of that country that may significantly affect the quality of the environment must be included in a detailed statement containing: the environmental impacts, the adverse effects that cannot be avoided, the alternatives for action, the relationship between the uses of environmental resources in the short term and the maintenance and improvement of their productivity in the long term, and any irreversible or irretrievable impairment of these resources if the proposal is implemented.

It is a document, part of the EIA process, which summarises the environmental and social conditions of the affected area in accordance with the guidelines already imposed by the environmental agency, which in turn has already assessed the impacts, presented mitigating measures and even opted for the environmentally viable locational alternative (IAIA, 2004).

In Brazil, according to Moreira (1992), the first piece of legislation referring specifically to the EIA is recorded in Federal Law No. 6.803 of 2 July 1980 (Industrial Zoning Law in Critical Pollution Areas), which provides in Article 10 that: "special studies of alternatives and impact assessments that make it possible to establish the reliability of the solution to be adopted" for "the implantation of zones of strictly industrial use that are intended for the location of petrochemical, chlorochemical and carbochemical poles, as well as nuclear installations and others defined by law" (from the aforementioned Article 10).

Subsequently, the EIA was disseminated through National Environmental Council (CONAMA) Resolution 001/86, which defines minimum criteria for the presentation of the study, very similar to the proposals of Gilpin (1995), IAIA (2004) and Sadler (1996). However, in Brazil, the EIA is similar in content and operationalisation. The EIA appears in the legislation, except in the National Environmental Policy[3] , as an integral part of the EIA, being the stage of carrying out the assessment of the environmental impacts of the proposal and the presentation of mitigation and monitoring. This can be identified by comparing

---

[3] In Federal Law No. 6.938 of 31 August 1981, in its article 9° , item III, the assessment of environmental impacts is called an instrument of environmental policy, but there is no description of its meaning or any differentiation from the EIA.

international and national guidelines. In Table 2.1, the first two columns describe the **guidelines** proposed by Sadler (1996) and IAIA (2004) **for public policy** and the third column describes the **national guidelines for preparing the EIA,** proposed in CONAMA Resolution 001/86.

Table 2.1 - Comparison between international guidelines for implementing EIA as a public policy and national guidelines for preparing EIAs

Guidelines for implementing and operating environmental impact assessment as a public policy		Guidelines for preparing EIAs (CONAMA Resolution 001/86)
Proposed by Sadler (1996)	**IAIA (1999)**	
0 author subdivides public policy into preliminary and detailed evaluation, where the first phase aims to: • Determine whether EIA is necessary and at what level of detail e; • The drafting of the terms of reference, through the Identification of main issues and impacts that need to be addressed. The detailed assessment is made up of • Impact analysis to identify, predict and assess the importance of potential risks, effects and consequences; • Specify measures to avoid, minimise and	For the author, the EIA process should include: • *Screening:* Determine whether the proposal should be subject to EIA, if so, at what level of detail; • *Scoping:* Identify the problems and impacts that are important and establish the terms of reference for the EIA; • *Examinationof alternatives:* establish the alternative environmentally viable; • *Impactanalysis :* identify and predict the impacts and effects associated with the proposal; • *Mitigation and impact management:*	0 EIA must: • Consider all the technological and localisation alternatives for the project, comparing them with the hypothesis of non-execution; • Systematically identify and assess the environmental impacts generated during the implementation and operation phases of the activity; • Define the boundaries of the geographical area to be directly or indirectly affected by the impacts; • Consider government plans and programmes proposed and being implemented in the project's area of influence and their compatibility. • And you'll still have

	determining the neces	
	sary measures for minimise or compensate	
compensate for environmental losses and damage; • Establishing the preferred or most environmentally error-free option in order to achieve the objectives of the proposal; • Determine the relative im portance and acce ptability residual impacts, i.e. those that cannot be mitigated; •Pres entation of reports wit h results of the environmental impact assessment, including the terms and conditions recommended in the EIS *(Environmental Impact Statement,* which in Brazil is called EIA/RIMA[4]);	the impacts , incorporating them into an environmental management plan; •*Ev aluationof significance:* Determine the relationship between the importance of impacts and the acceptability of residual impacts; •*Preparationof environmental impact statement (EIS) or report -* Clear and unbiased documentation of the impacts of the proposal, with mitigating measures, the importance of the proposal and the impact on the *environment.* impacts concerns of the public concerned and the communities affected by the proposal; •*Review of the EIS:* check that the report fulfils the terms of reference and	the following minimum technical activities: • Environmental diagnosis of the area of influence of the project with a complete description and analysis of the environmental resources and interactions, as they exist, in order to characterise the environmental situation of the area, prior to the implementation of the project, considering: the physical environment, the biological environment and the socio-economic environment; • Analysing the environmental impacts of the project and its alternatives, by identifying, predicting the magnitude and interpreting the impact on the environment. importance of likely impacts

---

• Review of the report, checking that it complies with the terms of reference; • Decision-making to approve (or not) a proposal and establish the rules and	provides an assessment satisfactory of the proposal, also containing the information necessary to take	relevant; • Definition of mitigation measures negative impacts , between them equipment control and waste treatment systems, evaluating the efficiency of
conditions; - 0 monitoring to check that the actions are in line with the terms and conditions and that the impacts are within the expected ranges, with the management of unforeseen events.	decisions; • *Decision-making* : approval or rejection of the proposal, with the establishment of terms and conditions for its implementation; • *Followup* : monitoring of the conditions established and verification of the effectiveness of the measures, in order to subsidise future EIA applications and contribute to monitoring the state of the environment and sustainable development.	each of them and drawing up the follow-up and monitoring programme, indicating the factors and parameters to be taken into account.

Source: Prepared by the author, adapted from Sadler (1996), IAIA (2004) and CONAMA Resolution 001/86.

With the exception of drawing up the terms of reference, public participation and decision-making, the Brazilian regulations include all the EIA guidelines. Public bodies fail to carry out a preliminary impact assessment, passing this responsibility on to the technical team hired by the developer. This may have been done on purpose, since environmental agencies, like most Brazilian public bodies, do not have the personnel, let alone the qualifications, to carry out the analysis. However, what can be seen is that in this context, **EIA in Brazil has come to be characterised as a technical activity to be carried out in the EIA and not as a public policy with broader and more global approaches, as was initially proposed internationally.** Furthermore, the EIA became subject to environmental licensing, becoming a technical document to be presented in the process of releasing a licence for a proposed project, no longer encompassing government plans and programmes.

2.3 Flaws in the preparation of Environmental Impact Studies in Brazil

What can be observed is that the inclusion of the EIA/RIMA as a command and control instrument has deficiencies identified in various countries, related to **operational and technical failures on the part of the public body** (MONTAZ, 2001; IBAMA, 1995; GOMES E AMÂNCIO, 2000) and **deficiencies in the preparation of the study itself**, where various characteristics cited by Gilpin (1995) are not assessed, or if assessed, are studied insufficiently.

The main shortcomings of EIAs/RIMAs were listed by Nicolaídes (2005) in her study on the effectiveness of Environmental Impact Assessment. Some of the weaknesses identified by the author are shown in Table 2.2:

Table 2.2 - Main shortcomings of the EIA in Brazil

Terms of Reference	Inadequately prepared; - It does not cover all the intrinsic variables of the proposed project.
Location Alternatives	• There is an absence or insufficiency in the proposal of locational alternatives; • A tendency for economic aspects to prevail over environmental ones, with the entrepreneur's preferred alternative prevailing; The analysis of impacts is carried out only for the chosen alternative and not for all possible alternatives, which would subsidise the multidisciplinary team's decision on the environmentally viable alternative.
Areas of Influence	- There is a disregard for the affected river basin, and the

	delimitations of the areas of influence have no basis in the characteristics and vulnerabilities of the natural environments and the social and regional realities that justify their confusing delimitation.
Environmental Diagnosis	• There is insufficient time to carry out field research, and the characterisation of the area is based predominantly on secondary data, with no or insufficient information on the method used; • There is a recurring proposal to carry out diagnostic activities after the Preliminary Licence (LP); • There is a lack of integration of data from specific studies, where the physical, biotic and anthropic environments are treated separately.
Environmental Impact Assessment	• It has been compromised due to failures in the previous stages, particularly in the diagnosis, with certain impacts not being identified; The Environmental Impact Assessment methods used in Brazil come, in most cases, from international experiences, with inadequate and/or precarious methodological procedures for predicting and interpreting impacts; • There is also a lack of evaluation of the synergistic and cumulative effects of the various impacts, including secondary impacts,
	indirect and cumulative that are not well identified and properly evaluated; • There is a lack of understanding that the EIA is a sequential process and, as such, the results of the EIA, such as the analysis of impacts, the proposal of measures and the environmental prognosis, do not, for the most part, obey a causal relationship with the diagnosis made; • There is a partial identification of impacts with the identification of generic impacts, making it impossible to understand the full extent of the changes expected from

	the project;
	• There is also the identification of mutually exclusive impacts and an under-utilisation or disregard of diagnostic data, despite there being sufficient information to identify impacts;
	• There is a lack of data and/or justification regarding the method used to assign weights to the elements of the impacts, since the studies do not always present and discuss the method used;
	• The studies are biased in that they minimise or underestimate the negative impacts and overestimate the positive ones;
	• Most environmental impact studies concentrate on carrying out exhaustive inventories, with little emphasis on the stage of identifying impacts.
Proposed measures	The specifications of the mitigation measures are precarious and do not reflect the solution for mitigating the impact;
	• Legal obligations are mentioned as mitigation measures, as if meeting them were an advantage;
	• There is no assessment of the technical effectiveness of the proposed mitigation measures.
Programmes monitoring	• There are recurrent conceptual errors in the indication of monitoring;
	• There is a lack of precision in the specifications and errors in not covering the entire area of influence, with the stipulation of execution deadlines that are incompatible with the occurrence of the impact.

Source: Prepared by the author, adapted from Nicolaides (2005).

The content of the EIA encourages decision-making on whether or not to implement a project. The focus of public policy is on the information presented in the study. Participants in the process, be they environmental agencies, public bodies, stakeholders or society in general, believe in the information presented and base their decisions on it. It is therefore essential that studies are based on credible scientific information that meets the proposed guidelines. Flaws such as those highlighted in Table 2.2 lead to wrong public decisions, preventing the development of environmental sustainability.

Analysing environmental impacts is the most complex action to be carried out as part of the EIA. By verifying and classifying the environmental elements of the impacts, it seeks to predict, identify and assess all the environmental changes caused by the implementation of the project, allowing for their correct mitigation, compensation and monitoring. Making it fault-free is not a very simple task.aNicolaídes (2005) in table 2.2 indicates several environmental elements that are addressed in the EIAs analysed, but the literature points to several other elements that should be studied, which are listed in section 2.4.

2.4 Environmental Reference Elements for Environmental Impact Assessment

The properties of environmental impacts are given different names and can be found in the literature as "criteria for assessing the importance of environmental impacts" (SANCHEZ, 2006; LAWRENCE, 2007; CEAA, 2009; ROSSOUW, 2003), "attributes of environmental impacts" (NICOLAÍDES, 2005) "reference parameters for environmental impacts" (IEEM, 2008) or even as "environmental elements" (NEPA, 1969). Due to the lack of homogenisation of concepts and the absence of a more specific denomination in Brazilian legislation, in this dissertation, in particular, we will use the expression **"environmental elements"** to relate the various properties of the impacts analysed, following the nomenclature adopted by American legislation. Although the term **"criterion"** is referenced more often in the literature, if it were adopted here, it could lead to conceptual confusion with the public policy criteria presented in Chapter I, which is undesirable.

Here we have two major questions to discuss: **1 - Which environmental elements should be analysed in an Environmental Impact Assessment? And 2 - How should each environmental element be conceptualised?**

Analysing the first question, we see that the American legislation "The National Environmental Policy Act of 1969 - NEPA", a pioneer in Environmental Impact Assessment, instituted in 1969, in its section 1502.16 entitled "Environmental consequences", establishes that when comparing alternatives, it must be verified:

- Direct effects of impacts and their significance;
- Indirect effects of impacts and their significance;
- Possible conflicts between the proposed action and compatibility with federal, regional, state and local plans;

  - Environmental effects of the alternative including the proposed action and comparisons between the alternatives;

- Potential energy consumption and conservation of the various alternatives and their mitigation actions;
- Depletion of natural resources;
- Urban, historical and cultural quality of resources, as well as resource utilisation, including reuse and conservation potential, and;
- Proposing mitigating measures for adverse impacts.

In the Brazilian legislation, which was based on the American one, these particularities are summarised as follows: *"identification, prediction of the magnitude and interpretation of the importance of the probable relevant impacts, breaking down: the positive and negative impacts (beneficial and adverse), direct and indirect, immediate and medium and long-term, temporary and permanent, their degree of reversibility, their cumulative and synergistic properties, the distribution of social burdens and benefits"* (CONAMA RESOLUTION No. 001/86).

In addition to the related legislation, some authors describe the environmental elements that should be part of the impact analysis, but without a more specific justification of why they were chosen, and there is still heterogeneity in these choices. To make it easier to understand the authors' approaches, the different elements chosen are listed in Table 2.3.

Table 2.3 - Environmental elements listed in scientific literature

ENVIRONMENTAL ELEMENT	BIBLIOGRAPHICAL REFERENCE									
	NEPA, 1969	CON AMA, 1986	IEEM, 2008	EPA, 2008	SANC HEZ, 2006	SANT OS, 2004	LAW RENC E, 2007	CEAA, 2008	PARR, 1999	RO HD E, 1988 apud Mor eira, 1992
Compatibility with government plans, projects and programmes	x	-	-	x	-	-	x	x	-	-
Influence	x	x	-	x	x	-	x	x	-	x
Distribution of social burdens and benefits	-	x	-	-	-	-	–	–	-	x
Duration (time scale)	x	x	x	-	x	-	x	x	x	x
Scope	-	-	x	-	-	-	x	x	x	x
Frequency	-	-	x	-	-	-	x	x	x	x
Importance	x	x		x	-	-	–	–	-	x
Magnitude	x	x	x	x	–	-	x	x	-	x
Nature (positive/negative)	x	x	x	x	x	-	x	x	-	x
Occurrence (probability)	x	-	x	x	x	-	x	x	-	-

Cumulative and synergistic properties	x	x	-	x	x	-	x	x	-	x
Reversibility	-	x	x	-	x	-	x	x	-	x
Density	-	-	-	-	-	-	-	-	x	-
Type	-	-	-	-	-	-	-	-	x	-
Trigger	-	-	-	-	-	-	-	-	-	x
Quantity	-	-	-	-	-	-	-	-	x	-
Configuration	-	-	-	-	-	-	-	-	x	-

Source: Prepared by the author.

The environmental elements predominantly studied are: influence, duration, magnitude, nature, occurrence, cumulative and synergistic properties and reversibility. It is interesting to note that the same organisation, in different publications, such as NEPA (1969) and the EPA (2008) present different environmental elements. It seems to us that the different approaches between the authors are due to the lack of a more detailed description within the legislation, which guides decisions, specifying and detailing in greater depth the environmental elements to be addressed. This lack of standardisation allows scholars to stipulate the environmental elements they consider most relevant and most comprehensive in relation to the characteristics of the impacts.

The empirical investigation used eight EIAs, submitted to the Brazilian Institute for the Environment and Renewable Resources (IBAMA) as a condition for licensing the activity. All of them are related to the electricity sector, more specifically: one for a hydroelectric dam, two for transmission lines, one for a radioactive waste dump, one for energy reinforcement, two for a gas pipeline and one for a gas treatment plant. There are five different preparation companies, each made up of a different multidisciplinary team. Even the EIAs drawn up by the same company differ in terms of team compared to the other studies.

In the studies analysed, the lack of standardisation is also very evident, and can be seen in the heterogeneity of the use of environmental elements, as shown in table 2.4.

Table 2.4 - Environmental elements found in the EIAs analysed

ENVIRONMENTAL ELEMENT	EIA'S ANALYSED							
	EIA 01	EIA 02	EIA 03	EIA 04	EIA 05	EIA 06	EIA 06	EIA 07
Compatibility with government plans, projects and programmes	-	-	-	-	-	-	-	-
Influence (direct/indirect)	-	x	x	x	x	-	-	x
Distribution of social burdens and benefits	-	-	-	-	-	-	-	-
Duration	x	-	x	x	x	x	-	x
Time scale or temporality	-	-	x	x	x	x	-	x
Scope	x	x	x	x	x	x	x	x
Frequency	-	-	-	-	-	-	-	-

Importance	-	-	X	X	X	-	X	-
Magnitude	-	X	X	X	X	X	X	-
Nature (positive/negative)	X	X	X	X	X	X	-	-
Occurrence (probability)	-	X	-	-	-	X	X	-
Cumulative and synergistic properties	-	-	-	-	-	-	-	-
Reversibility	X	X	X	X	X	-	X	X
Significance	-	-	X	X	X	-	-	X

Source: Prepared by the author.

Table 2.4 shows that the technical team has a different way of approaching the environmental elements compared to what is found in the literature. There is a maintenance of the elements: influence, duration, magnitude and reversibility; an addition of temporality and extension; and an omission of the environmental elements: nature, occurrence and synergistic properties, predominant in scientific literature.

In addition to the problem related to the definition of the environmental elements to be addressed, another relevant question under discussion here is: **how do you conceptualise each environmental element?** In order to truly analyse the environmental elements, it is necessary to be clear about what each one means, which properties of the impacts are addressed in each element, how they are classified, whether there is a ranking of importance or even the interrelationships between the various elements studied.

What we see, however, is that each author or technical team has their own way of conceptualising these elements and their interactions. Most of the time, these approaches are summarised and subjective. The authors only present a classification of the environmental element, without a justification for the categorisation, which seems to be done subjectively.

In the case of the EIAs analysed here, there is also a particularity: the apparent homogenisation of concepts, classifications and definitions is due to the fact that different studies were carried out by the same company or the same team, and not necessarily a coherence of thought on the part of the authors. Some of the definitions found in the literature and in the studies are summarised in tables 2.5 and 2.6.

Table 2.5 - Concepts found in scientific literature

Elements Environmental	Author	Meaning
Compatibility with government plans,	NEPA, 1969; EPA, 2008;	The authors don't give a conceptualisation, they just explain that the team preparing the

projects and programmes	LAWRENCE, 2007; CEAA, 2008.	EIA must check the impacts of the activity in relation to its existing regulations and local public policies, verifying that it is compatible and legal.
Distribution of burdens and bonuses	ROHDE, 1988 apud Moreira, 1992	The author doesn't present a concept, he just classifies them as socialised and privatised.
Trigger	ROHDE, 1988 apud Moreira, 1992	There is no defined conceptualisation, only a classification, which can be considered immediate, differentiated or staggered.
Type	PARR, 1999	It conceptualises the element as the reference to the disturbance generated, which can be: similar or different impacts.
Quantity	PARR, 1999	It also refers to the disturbance generated by the impact, classified as single or multiple.
Configuration	PARR, 1999	It refers to the spatial characteristics of the impact and can be classified as punctual, linear or in area.
Density	PARR, 1999	Refers to the spatial attributes of the impact, and can be differentiated between clusters and separate ones
Magnitude	ROHDE, 1988 apud Moreira, 1992	It only classifies it as large, medium or small, without detailing a concept about the element.
	IEEM, 2008	Refers to the size or quantity of an impact, determined on a quantitative basis if possible.
Occurrence	IEEM, 2008	Subdivided into 1 - certain or almost certain, when the estimated probability of occurrence is 95% chance or higher; 2 - probable, when the estimated probability is above 50% but less than 95%; 3 - unlikely, when the probability is greater than 50%.
		estimated to be above 5 per cent, but less than 50 per cent, and; 4 - extremely unlikely, when

		the estimated probability is less than 5 per cent.
	SANCHEZ, 2006	It refers to the degree of uncertainty about the occurrence of an impact; it can be classified according to the following scale: 1 - certain, when there is no uncertainty about the occurrence of the impact; 2 - high, when based on similar cases and observation of similar projects, it is estimated that it is very likely that the impact will occur; 3 - medium, when it is unlikely that the impact will occur, but its occurrence cannot be ruled out and 4 - low, when it is very unlikely that the impact in question will occur, but even so this possibility cannot be dismissed.
Nature	IEEM, 2008	They are defined only as those impacts that bring positive or negative results to biodiversity
	SANCHEZ, 2006	Defined by the author as "expression", it describes the positive or negative character of each impact; although most impacts have a clear positive or negative character, some impacts can be both positive and negative, i.e. positive for a certain component or natural element and negative for others.
Extension	IEEM, 2008	For the author, the extent of an impact is the area in which the impact occurs.
	ROHDE, 1988 apud Moreira, 1992	Classified only as punctual, extensive (in area), linear or spatial.
	PARR, 1999	This refers to the spatial attributes of the impacts, which can be classified as local, regional or global.
	SANCHEZ, 2006	Referred to by the author as the spatial scale, it is defined as: 1 - local impacts, those whose scope is restricted to the limits of the project areas; 2 - linear impact is that which manifests

		itself along the roads transporting inputs or products; 3 - municipal scope is used for impacts whose area of influence is related to municipal administrative limits; 4 - regional scale is used for impacts whose area of influence goes beyond the two previous categories, and may include the entire national territory; and 5 - global scale, for impacts that potentially affect the entire planet.
Duration	PARR, 1999	It conceptualises the time scale as the temporal characteristics of the impact, which can be differentiated into the short term and the long term.
	IEEM, 2008	According to the author, the duration of the impact can be defined as the time in which the impact is expected to occur. This
		should be defined in relation to ecological characteristics rather than counting days per se.
	ROHDE, 1988 apud Moreira, 1992	0 author only classifies it as: 1 year or less, 1 to 10 years and 10 to 50 years, without presenting a more specific concept.
	SANCHEZ, 2006	They are classified as: 1 - temporary impacts, which are those that only manifest themselves during one or more phases of the project and which cease when the project is decommissioned. These are impacts that cease when the action that caused them ends; and 2 - permanent impacts, when they represent a definitive alteration to a component of the environment or, for practical purposes, an alteration that lasts indefinitely, such as the degradation of soil quality caused by waterproofing due to the construction of a shopping centre and car park; these are

		impacts that remain after the action that caused them ceases.
Reversibility	IEEM, 2008	For the author, an irreversible impact is one for which recovery is not possible within a reasonable time or for which there is no reasonable possibility of action being taken to reverse the fact; a reversible impact occurs when spontaneous recovery is possible, or there are effective means of mitigating the impact. According to the author, in some cases the same activity can cause both reversible and irreversible impacts.
	ROHDE, 1988 apud Moreira, 1992	It is only considered reversible/temporary or irreversible/permanent, without an explanation of their concept.
	SANCHEZ, 2006	According to the author, the concept of reversibility is the system's ability to return to its previous state if the external demand ceases or if a corrective action is implemented.
Frequency	IEEM, 2008	The author defines frequency as when changes can only have an impact if they coincide with critical life stages or seasons. The frequency of an activity, and therefore the impact resulting from it, must also be considered.
	ROHDE, 1988 apud Moreira, 1992	It only presents a classification of the element, which can be: continuous, discontinuous or dependent on the time of year.
	PARR, 1999	It only classifies it as: continuous and discontinuous.
Importance	IEEM, 2008	According to the author, the importance of biodiversity must be specific and the integrity of habitats and the state of conservation must be verified.

	ROHDE, 1988 apud Moreira, 1992	0 author only presents a classification, without detailing the concept. 0 impact can be classified as: important, moderate, weak, negligible.
	NEPA, 1969	In order to establish the importance of an impact, the following must be observed: 1 - Context: this means that the significance of an action must be analysed in various contexts in society as a whole (human, national). The region affected, the interests affected and the locality must be analysed. The significance of the impact varies according to the definition of the proposed action; 2 - Intensity: this refers to the severity of the impact.
Influence	SANCHEZ, 2006	For the author, it is the cause or source of the impact, whether direct or indirect; 1 - direct impacts are those that result from the activities or actions carried out by the developer, by companies contracted by them, or that they can control; 2 - indirect impacts are those that result from a direct impact caused by the project itself, in other words, they are second or third order impacts; indirect impacts are more diffuse than direct impacts and manifest themselves in wider geographical areas.
	ROHDE, 1988 apud Moreira, 1992	Only categorised as direct, when they come from primary effects, and indirect, if they come from secondary, tertiary effects, etc.
	NEPA, 1969	They are classified as: 1 - direct effects when they are caused by the action and occur at the same time and place; 2 - indirect effects when they are caused by the action of time, observed after the primary impacts, but are still reasonably foreseeable.

Cumulative properties and synergistic	SANCHEZ, 2006	Cumulative impacts are those that accumulate in time or space and result from a combination of effects arising from one or more actions.
	CANTER, 1995	According to the author, cumulative impacts refer to their environmental characteristics in terms of their association with other environmental impacts and the potentialisation of changes caused by future actions.
	ROHDE, 1988 apud Moreira, 1992	Only one classification is presented, in the case of cumulative impacts, they can be classified as: linear, quadratic, exponential, etc. And for synergy, whether it is present, yes or no.
	NEPA, 1969	It is conceptualised by the author as the impact on the environment that results from the incremental impact of the action when added to other past, present and future actions
		reasonably foreseeable. Cumulative impacts can be individually smaller than when significant collective actions occur.

Source: Prepared by the author.

What we see is that most authors don't even present a conceptualisation, they just present a classification, which is also different for each author, some apparently similar, others quite different. There is a confusion of terminology, especially between magnitude, significance and importance, making it difficult to understand and conceptualise the terms.

Checking the operationalisation of the EIA in an empirical investigation, we observed that there is also a difference in terminology and conceptualisation in the EIAs under analysis, as can be seen in table 2.6.

Table 2.6 - Concepts found in the EIAs studied

Elements Environmental	Author	Meaning
Magnitude	EIA 06	It quantifies the effects, which can be small, medium or large.

	EIA 01	Defined by the author as the intensity of the impact, it refers to the degree of incidence of an impact on an environmental factor, in relation to its universe, as it is present in the area of influence, identified as: strong, medium, weak and variable intensity. The authors assign scores for each classification, from 10 to 8 for strong, 7 to 4 for medium and 3 to 1 for weak intensity.
	EIA 03 / EIA 04 / FIA 05	It refers to the degree of incidence of an impact on the environmental factor, in relation to its universe. According to the team preparing the study, the magnitude is related to the size of the impact, which can be large, medium or small, according to the
Nature		intensity of transformation of the pre-existing situation of the impacted environmental factor.
	EIA 02 / EIA 07	Magnitude refers to the degree of impact on a specific environmental parameter and in relation to that environmental factor as a whole. It can be high, medium, low or insignificant, depending on the intensity with which the environmental factor is modified. Considering that the impact may occur, it is then assessed regardless of the likelihood of its occurrence. According to the multidisciplinary team, the magnitude of the impact is classified exclusively by the relationship between the environmental factor in question and the activity, i.e. it does not take into account the possibility of affecting other environmental factors.
Occurrence	EIA 02 / EIA 07	According to the study team, the probability or frequency of an impact will be high if its occurrence is almost certain and constant throughout the activity; medium if its occurrence is intermittent and low if it is practically unlikely to occur.
Nature	EIA 08	For the team, this element represents the influence of an action carried out on the project, resulting in a non-significant positive change in the area. It can be

		classified as: 1 - significant positive impact (when an action carried out on the project results in a significant positive change in the area); 2 - non-significant positive impact (when an action carried out on the project results in a non-significant positive change in the area); 3 - negative impact
		significant (when an action carried out in the undertaking results in a significant negative change in the area); 4 - non-significant negative impact (when an action carried out in the undertaking results in a non-significant negative change in the area) and 5 - undefined impact (when an action carried out results in an environmental change that is still uncertain, as it depends on the tools, methods and intensity used in the impacting action, becoming positive or negative through mitigating measures).
	EIA 06	They only characterise them as negative and positive effects, without conceptualising them.
	EIA 01 / EIA 02 / EIA 03 / EIA 04 / EIA 05 / EIA 07	It is conceptualised as the nature of the impact. It indicates whether the impact produces beneficial/positive or adverse/negative effects on the environment.
Extension	EIA 08	It refers to the spatial delimitation of the impact based on the reduction between the causative action and the territorial extension affected. It is classified as 1- local (when the extent of the impact reaches the surface delimited by the area of direct influence and a small peripheral portion of the land); 2 - regional (when the extent of the impact reaches the surface delimited by the area of functional influence and its hydrographic basin); 3 - strategic (when the extent of the impact occurs in a strategic policy).
	FIA 05	Indicate the impacts whose effects are felt

		locally, in the immediate vicinity of the activity, or that can affect wider geographical areas; regionally, or when they have a strategic characteristic, with national scope.
	EIA 02 / EIA 03 / EIA 07	Indicates impacts whose effects are felt locally, in the immediate vicinity of the activity, or which can affect wider geographical areas, regional. Broad impacts on ecosystems have been classified as regional.
	EIA 01	Defined as the extent of the area covered by the manifestation of the effects of the impacts and classified as local, regional and global.
	EIA 06	Situates the scope of the impact, whether it is localised or dispersed
Duration	EIA 08	Described as temporality, it represents the temporal form of occurrence of the environmental impact, presenting itself in a dimension that becomes gradual to the different actions that produce impacts on the environmental system, and can be classified as: 1 - temporary (when the impacting factors cease after the generating action is interrupted) and 2 - permanent (when the impacting factors remain after the generating action is interrupted).
	EIA 01 / EIA 05	They only divide the impacts into permanent, temporary or cyclical, i.e. those whose effects manifest themselves indefinitely, over a set period of time or cyclically, and can occur seasonally.
	EIA 03 / EIA 04 / EIA 06	It indicates how long the impact will last, and can be categorised as temporary or permanent.
Reversibility	EIA 08	It refers to the ability of the element of the environment affected by a particular action to return to its previous environmental conditions. It is classified as 1 - reversible (when after an impacting action the affected environmental object returns to the initial environmental conditions, either naturally or anthropically); 2 -

		irreversible (when the environmental object affected by an impacting action does not reach previous environmental conditions, despite attempts to do so).
	EIA 01 / EIA 02 / EIA 03 / EIA 07 / EIA 05	It classifies impacts according to whether their effects are irreversible or reversible. It makes it possible to identify which impacts can be completely avoided or which can only be mitigated or compensated for.
Time scale	EIA 08	This can be: 1 - immediate (when the neutralisation of the impact occurs after the end of the action); 2 - medium-term occurrence (when there is a need for a reasonable period of time to elapse for the impact to dissolve); and 3 - long-term occurrence (when after the conclusion of the action generating the impact, it remains for a long period of time).
	EIA 05	It differentiates between impacts that manifest themselves immediately after an impacting action, in the short term, and those whose effects are only felt after a period of time has passed in relation to their cause, in the short, medium or long term.
	FIA 04 / EIA 03	It differentiates the impacts according to manifested immediately after the impacting action, in the short term, and those whose effects are only felt after a period of time has passed in relation to their cause.
	EIA 06	It indicates when the impact occurs, and can be short, medium or long term.
Importance	EIA 02 / EIA 07	Significance is associated with the degree of interference that specific actions or operational processes can have on the different environmental parameters. It takes into account not only the magnitude of the impact, but also its probability of occurrence. A potential impact can be of potentially high magnitude with a low probability of occurrence, leading to a low significance. It can thus have the

		following classifications: High, medium, low or insignificant, according to the degree of interference with environmental factors.
	EIA 03 / EIA 04 / EIA 05	It refers to the degree of interference of the environmental impact on different environmental factors, and is strictly related to the relevance of the environmental loss, which can be large, medium or small, to the extent that it has a greater or lesser influence on the local environmental quality as a whole.
Influence	EIA 08	Called spatialisation by the authors, it is the attribute by which the level of relationship between the impacting action and the impact generated on the environment is determined, and can be classified as 1 - direct (also called primary or first-order impact. It results from the actions of the undertaking on the elements of the environment); 2 - indirect (results from a secondary action in
		response to the previous action or when it is part of a chain of reactions (also known as a secondary or nth order impact).
	EIA 02 / EIA 03 / EIA 04 / EIA 05 / EIA 07	It is how the impact manifests itself, i.e. whether it is direct (resulting from an action carried out by the project) or indirect (resulting from an accident or unexpected occurrence, or a secondary impact caused by the main impact.

Source: Prepared by the author.

The technical teams present scales of values for the environmental elements that have no scientific basis, and confuse terms such as magnitude and importance. Most of the teams have a similar understanding of the subject, probably due to the informal "standardisation" of the EIAs prepared in Brazil, listing the same environmental elements, differing only in their classification.

In addition to the gaps related to which environmental elements should be addressed and the correct definition for these elements, another obstacle that should be investigated is the methods used to highlight these elements and classify them. According to Lawrence (2007) and Rossouw (2003), environmental elements should be classified using environmental impact methods, with environmental elements being listed, which may contain different levels of scale, progressing from the smallest to the most complex, from quantitative to qualitative and from individual to cumulative. The methods used for demonstration and evaluation will be discussed in Chapter III.

CHAPTER 3

ENVIRONMENTAL IMPACT ASSESSMENT METHODS

3.1 Origin and evolution of Environmental Impact Assessment methods

Environmental Impact Assessment methods emerged in response to the guidelines imposed by the National Environmental Policy Act (NEPA), first with the checklist, matrices, interaction networks and *input-output* methods (GILPIN, 1995).

Since 1976, EIA methods have reflected a number of scientific advances aimed at solving the specific problems of each environmental impact study. Although the pioneering EIA methods continued to be used, there was an evolution towards a better understanding of the cause and effect relationships between project actions and their impacts, and towards taking into account the dynamics of environmental systems. In the 1980s, the conceptual basis for the scientific approach to environmental impact assessment began to emerge (MOREIRA, 1992).

According to Moreira (1992), the evolution of environmental policy administrative procedures and the participation of social groups in the discussion process have standardised procedures, mainly in relation to: 1 - defining the scope of studies by means of terms of reference, determining the relevant environmental factors and the fundamental issues for decision-making; 2 - inserting procedures for monitoring and reviewing studies; 3 - defining legally established environmental quality standards that contribute to the emergence of new methods (there are several studies related to specific methods for assessing each impact and effects resulting from specific activities).

3.2 Main methods of Environmental Impact Assessment

One of the main objectives of Environmental Impact Assessment is to predict the changes resulting from the implementation of a project. To this end, various methods are available for identifying, assessing and predicting impacts (MOREIRA, 1992; SANCHEZ, 2006; PARR, 1999).

The scientific framework on the subject is limited. There are several limitations in the available literature, the main ones being: 1 - The authors are not aligned on the classification of the methods (whether they are for prediction, identification or evaluation), and most are silent on the subject; 2 - The authors describe the same methods under different names, making it difficult to understand the subject and; 3 - There is confusion between traditional environmental impact assessment methods and methods and techniques that are used to complement the assessment.

The diversity of classifications can be seen in Parr (1999), Moreira (1992) and Sanchez (2006). In the European Commission report entitled *"Study on the assessment of indirect and cumulative impacts as well as impact interactions"*, prepared by Parr (1999), the methods are divided into two groups: the first group is made up of **predictive methods** (used during impact prediction and identification); and the second group is made up of methods used to assess the significance of impacts, known as **assessment methods.** The main differences between the classifications are summarised in Table 3.1.

In Sanchez (2006) we see the following subdivision: 1 - **methods for prospecting impacts** (hypothesis about the future behaviour of some environmental parameters that informs the magnitude or intensity of environmental modifications); 2 - **methods** for **identifying** impacts (aims to identify probable impacts, facilitating understanding of the environmental elements and processes that may be altered by the project) and; 3 - **methods for assessing** the importance of impacts (discusses the importance or significance of impacts, defined by means of a value judgement). The environmental impact assessment **techniques** found in Moreira (1992) define the specific operations of discovering facts or manipulating information, data or knowledge and are designed to estimate the magnitude of the impacts that will be caused by the actions to be carried out when the project is realised.

The methods observed in the literature are listed in table 3.1 and reflect the heterogeneity of the names used by the authors, but by studying the definition of each method presented, it was possible to establish a connection between the different terms. These methods will be summarised in section 3.3.

Table 3.1 shows the differentiations between the classifications of environmental impact assessment methods and the confusion between assessment methods, evaluation techniques and valuation techniques, already demonstrating that the literature on the subject has several important gaps.

Table 3.1 - Survey of Environmental Impact Assessment methods in the literature

Classification	Reference	Method
Forecasting methods	Parr, 1999	• Checklists; • Matrices; • Mathematical models; • Interaction networks;

		• Card overlays.
Impact prospecting methods	Sanchez, 2006	• Mathematical models; • Comparison and extrapolation; • Laboratory and field experiments; • Simulation and analogue models; • Mass balance; • Expert judgement.
Methods for identifying impacts	Sanchez, 2006	• Checklists; • Ad hoc; • Leopold Matrix; • Interaction matrix.
Environmental impact assessment methods	Parr, 1999.	• Cost-benefit analysis; • Multi-criteria method.
	Moreira, 1992	. Ad hoc; • Checklists; • Interaction matrices; • Interaction networks; • Superimposition of cards; • Simulation models.
	Braga, 2006	• *Ad hoc;* • Checklists; • Card overlays; • Interaction networks; • Interaction matrices (Leopold); • Simulation models; • Cost-benefit analysis; • Multi-objective analysis.
	Gilpin, 1995	. Checklist; • Matrices; • Cost-Benefit Analysis (CBA); • Cost Effectiveness Analysis (ACE); • Opportunity Cost; • Multiplier method; • Contingent Valuation Method (CVM); • Travel Cost Method (TCM); • Hedonic Price Method (MPH)

		. Checklist;
	Santos, 2004	• Matrices;
		• Spatial analysis;
		• Enquiry;
		. Decision trees;
		• Modelled and simulation systems.
Methods for assessing importance	Sanchez, 2006	• Combination of attributes;
		• Weighting of attributes;
		• Multi-criteria method.
Environmental impact assessment techniques	Moreira, 1992	. Statistical projections.
		• Field and laboratory experiments.
		• Mass balance.

Source: Prepared by the author.

There is no standardised differentiation, what is understood is that methods that provide information only to identify the environmental impact should be called: **environmental impact identification method;** and those that allow an assessment of the issue are called **environmental impact assessment methods.**

Regardless of the classification of this method, defining which one should be used is the responsibility of the multidisciplinary technical team and depends on a careful analysis of the type of project and the information to be collected (MOURA, n.d.; SANTOS, 2004; BRAGA, 2006; HORBERRY, 1984). The report by the European Commission (1999) presented a study of sixty EIAs and found major deficiencies in the use of the methods. Analysing the main elements to be addressed, it noted the failure to identify indirect and cumulative impacts and the interactions between impacts.

According to Parr (1999), the application of these methods is limited in practice. No one method can be considered the best, nor is there a tool that can be used to assess all the stages of the study or that is suitable for assessing any type of project (LENSEN *et al,* 2003). It is widely accepted that a single method is not capable of bringing together all the environmental elements required for the effective assessment of impacts, especially indirect and cumulative impacts (PARDO, 1997; LINGHJEM *et al,* 2007; PARR, 1999).

Traditional methods have an inherent subjectivity in the process of identifying and classifying elements[5] and parameters. This observation was raised by most of the authors

---

[5] The environmental elements to be addressed in the Environmental Impact Assessment are listed in CONAMA Resolution No. 001/86, and are: prediction of the magnitude and interpretation of the importance of the probable relevant impacts, breaking down: positive and negative impacts (beneficial and

studied (MOREIRA, 1992; BRAGA, 2006; SANTOS, 2004; SANCHEZ, 2006; PARR, 1999; MOURA, n.d.). This subjectivity is detrimental to the process of analysing environmental impacts, since the analyst makes value judgements according to their moral, ethical and technical concepts, which, according to the authors, is the greatest weakness in the use of methods.

In an attempt to reduce subjectivity by minimising the insertion of value judgements through the introduction of monetary values to environmental impacts, Gilpin (1995) presents the following methods for Environmental Impact Assessment: Cost Benefit Analysis (CBA), Cost Effectiveness Analysis (CEA), Opportunity Cost, Multiplier Effect, Contingent Valuation Method (CVM), Travel Cost Method (TCM), Hedonic Price Method (HPM) and Ecological Valuation, demonstrating a clear conceptual error in relation to valuation methods and the objectives of Environmental Impact Assessment.

Environmental valuation methods are defined as analytical tools that make it possible to weigh up the different economic incentives that affect agents' decisions regarding the use of natural resources. In fact, they contribute to a more comprehensive form of project evaluation, being useful as an **auxiliary** tool for evaluating environmental impacts, allowing the different economic incentives to be identified and weighted (MOTA, 2001; PUGAS, 2006; YOUNG *et al*, 1997; NOGUEIRA, 1998), but they do not have the characteristic of identifying, evaluating and predicting impacts. For Hufschimidt *et al* (1983 *apud* NOGUEIRA, 1998), valuation should be carried out after assessing the physical, chemical and biological effects of the activities, i.e. after drawing up the EIA with its respective Environmental Impact Assessment.

Most methods of valuing environmental services that do not have market prices have been associated with the microeconomic theory of well-being, through the development of new methods of social cost-benefit analysis. These valuation methods seek to capture people's preferences for environmental goods, so the decision by individuals to pay monetary values for certain goods and not for others is based on individual preferences and the quest to maximise individual well-being (MOTA, 2001; PUGAS, 2006). As with traditional EIA methods, these monetary values assigned to environmental goods depend on a value judgement, making valuation subjective as well. However, the use of these methods could not reduce the subjectivity of the EIA process as proposed by Gilpin (1995).

---

adverse), direct and indirect, immediate and in the medium and long term, temporary and permanent, their degree of reversibility, their cumulative and synergistic properties and the distribution of the permanent social burdens and benefits, their degree of reversibility, their cumulative and synergistic properties and the distribution of the social burdens and benefits.

The **environmental impact assessment techniques** that appear in international literature are developed to meet specific objectives and conditions, seeking to delve deeper into the subject and include new environmental elements and indicators. These include the Methods for Cumulative Effects Assessment (CEA), the Rapid Impact Assessment Matrix (RIAM) and the Analysis of Ecological Risks (AER) method.

What is expected of an **environmental impact assessment method** is that it is able to predict, identify and analyse potential changes. The design of the method to be used in a given study must take into account the technical and financial resources available, the time it will take, the data and information that exists or can be obtained, the legal requirements and the terms of reference to be met (MOREIRA, 1992). Sanchez (2006) points to the interaction between tools and procedures as the main challenge in the practice of EIA.

3.3 Description and application of Environmental Impact Assessment methods

According to Braga (2006), what the methods have in common is that they discipline the reasoning and procedures aimed at identifying the causal agents and the changes resulting from an action or set of actions. Among the methods mentioned above, some are more widely used and discussed in academic circles. The following seeks to standardise the information provided by the various authors, describing the most relevant environmental impact assessment methods in national and international literature.

3.3.1 Environmental impact assessment methods

- Ad Hoc" method

Ad hoc" methods consist of setting up working groups made up of professionals and scientists from different disciplines, according to the characteristics of the project to be assessed. It is based on the ability of certain specialists to issue estimates on the probability of occurrence, spatial and temporal extent and magnitude according to the experience and knowledge of the scientists. It is a method that can be used when there is little time to analyse the impacts and a lack of data for their systematic treatment (MOREIRA, 1992; SANCHEZ, 2006).

- Checklists, Checklists

Of all the methods, checklists survive as a guide to the potential impacts of a project, demonstrating in preliminary analyses the identification of impacts, listing the environmental elements and their respective indicators. The lists can be simple, descriptive, scalar and weighted scalar. The scalar list allows values to be assigned to environmental factors, making it possible to order or classify them according to pre-established criteria. If the factors are given a weight, expressing the importance of the impact, then the list is called a weighted scalar list (GILPIN, 1995; MOREIRA, 1992; SANTOS, 2004).

Although widely available in the literature, it is unlikely that a checklist will be used without some adaptation, either due to the characteristics of the project or the conditions of the environment. What we see is that these lists do not correlate impacts and their causes, omitting cumulative effects from the identification of impacts, thus compromising the analysis. On the other hand, checklists make it easy to systematise information, with the ability to summarise results, speed of application and low cost (SANCHEZ, 2006; SANTOS, 2004; PARR, 1999).

- Matrices (Interaction Matrices, Leopold Matrix)

One of the criteria to be assessed in environmental impacts is the magnitude, which can be identified through the use of matrices, which correlate the main activities or actions that make up the project and the main elements of the environmental system, identifying the possible interactions between the components of the project and the elements of the environment (GILPIN, 1995; SANCHEZ, 2006).

The best known and most widely used is the Leopold Matrix, developed by Leopold at the United States Geological Survey (USA) in 1971. The matrix cross-references 100 actions with 88 environmental components, resulting in 8,800 intersection cells. The two elements of environmental impacts, magnitude and importance, are used to describe the interactions. Magnitude is the measure of the extent of the impact and importance the measure of its relevance and the environmental factor affected, compared to other impacts and the environmental characteristics of the affected area. Each cell representing a possible impact is marked with a diagonal line. At the top of the diagonal, the value of the magnitude attributed to the impact is noted, using a scale from 1 (lowest magnitude) to 10 (highest magnitude), identifying a positive impact with a + sign and a negative impact with a - sign; at the bottom, the value of the impact's importance is noted (GILPIN, 1995; MOREIRA, 1992; SANCHEZ, 2006; SANTOS, 2004).

According to Gilpin (1995), the Leopold Matrix has several disadvantages: 1 - it is impossible to distinguish between a highly probable but low magnitude impact and a catastrophic but low probability impact; 2 - the time horizon of the impacts is not revealed; 3 - several alternatives cannot be compared in a single matrix; 4 - there are no criteria for measuring magnitude and importance; 5 - there is no presentation of important secondary impacts and; 6 - there is a tendency to neglect social and economic values.

The most serious flaw is that the Leopold Matrix depends on the subjective assessment of the team and the judgement is converted to simple numbers, losing much of its analytical content, and there is also a great danger in analysing by trying to count the numbers in order to achieve the overall effect. Furthermore, its counterparts represent the environment as a set of compartments that are not related (SANCHEZ, 2006; SANTOS, 2004). The matrix should be used as a reference checklist or as a reminder of the broad spectrum of environmental actions and impacts, with the function of communication between the project team, readers and analysts (GILPIN, 1999; SANCHEZ, 2006).

- Interaction networks or decision trees

The interaction network is the method that best identifies cumulative impacts and effects and their interactions (PARR, 1999). Using diagrams, they organise discussions and the exchange of information on the impacts and interactions of environmental factors. Despite this, interaction networks should only be used to identify indirect impacts, as they do not highlight the relative importance of the impacts identified, nor do they dispense with the use of forecasting and other methods to complete the other tasks of the study (MOREIRA, 1992; GILPIN, 1995).

Sporbeck (1997 *apud* Parr, 1999) states that the method was developed for road projects and focuses on the ecosystem and landscape, differentiating three elements of impact interaction: *"ecosystematic interactions, impact-upon ecosystematic interactions and impact shift"*. The method allows for the identification of direct impacts on primary receptors, but also the monitoring of impacts on other elements of the ecosystem, resulting from interactions. The complexity of this method is its main disadvantage, acting as a barrier to its use in small-scale projects (PARR, 1999).

- Map Overlay, Spatial Analysis or Geographic Information System

EIA methods of the letter overlay type generally consist of drawing up a set of letters of the area to be affected on transparent material, individually representing the relevant environmental components (soil types, vegetation cover, drainage, etc). The least restrictive or most suitable areas for the development of the proposed project are marked in white and the most restrictive or completely unsuitable areas in black. The overlapping of thematic maps shows, in the lighter regions of the map thus produced, the areas where the impacts of the project would be minimal (MOREIRA, 1999; SANTOS, 2004; PARR, 1999).

According to Gontier (2006), the use of Geographic Information Systems (GIS) as a tool for predicting impacts has certain limitations that need to be taken into account. These include: 1 - the impossibility of quantifying impacts taking into account ecosystem interactions; 2 - the impossibility of including some important environmental factors that cannot be mapped, such as biodiversity species and ecological processes and; 3 - the difficult integration with socio-economic impacts. In addition, there is a lack of knowledge about the real response that biodiversity gives to the insertion of infrastructure components and other developments (MOREIRA, 1992; PARR, 1999).

- Simulation Models, Modelled Systems, Quantitative Methods or Mathematical Models

These methods aim to represent, as closely as possible to reality, the structure and functioning of environmental systems, exploring the relationships between their physical, biological and socio-economic factors (BRAGA, 2006).

The basic structure of a simulation model involves carrying out the following tasks: 1 - definition of the results to be obtained and choice of the relevant environmental factors and elements; 2 - limits of the project's area of influence; 3 - simulation time horizons; 4 - listing of the project's actions and possible alternatives; 5 - selection and organisation of the variables designed to describe the environmental factors relevant to characterising the system; 6 - construction of a flow diagram or interaction network between the variables and the subsystems, indicating the respective interaction rules; 7 - identification of the impact indicators for each variable; 8 - choice of computer programme and processing language; 9 - operation of the simulation model and; 10 - interpretation and discussion of the model results, further processing until the result is considered valid (MOREIRA, 1992).

Gontier (2006) points out that evaluations based on simulation models have the advantage of being based on real data, which locally makes the results relevant. However, this data is not always available or static over time, so modelling fails to produce real effects and extrapolating the data can produce misleading results.

3.3.2 Environmental Impact Assessment Techniques

■ Multicriteria Analysis

The multi-criteria method has been used with great frequency to solve environmental problems (MARTIN *et al*, 2007). They make it possible to evaluate criteria that cannot be transformed into financial values. Their application is appropriate for comparing project alternatives, policies and courses of action, as well as for analysing specific projects, identifying their degree of overall impact, the most effective actions and those that should be modified (VILA BOAS, n.d.).

The method is developed in stages, as follows: 1 - formulation of the problem; 2 - determination of a set of potential actions (alternatives that meet the problem); 3 - development of a coherent family of criteria to assess the effects caused by the action on the environment; 4 - assessment of the environmental elements, with the construction of an assessment matrix (actions to be assessed x elements); 5 - determination of the weights of the elements and discrimination limits, where the weights reflect the importance of each element and; 6 - aggregation of the elements, associating the matrix with a mathematical model (VILA BOAS, n.d.).

■ Rapid Impact Assessment Matrix (RIAM)

The Rapid Impact Assessment Matrix (RIAM) is a tool for carrying out an environmental impact assessment (EIA) used to organise, analyse and present the results of an EIA in a holistic way. RIAM was originally developed to compare different alternatives for a single project, plan or programme (KIUTUNEN *et al*, 2008). The method seeks to reduce the subjectivity of other methods by inputting quantitative data. The data can be entered into computerised systems and shown in graphs, offering various impact scenarios to be assessed, reducing the time spent carrying out the EIA (PASTAKIA and JENSEN, 1998).

According to Kiutunen (2008) the use of the method does not exclude conventional methods; on the contrary, these methods combined with RIAM provide a greater number of variables, enriching the assessment of environmental impacts.

■ Analysis of Ecological Risks (AER) method

The risk analysis method is suitable for identifying alternatives and assessing their environmental compatibility. Its transparency provides very useful results for the political "decision-maker". The main objective is a cause and effect analysis relating the impacts between the project and the ecosystem. The method considers the ecological state of the environment before and after the project.

The approach is based on geographical information, combining it with other existing assessment methods (SANKOH, 1996).

■ Methods for Cumulative Effects Assessment (CEA)

Cumulative effects assessment methods are an interrelationship with existing methods, which seek to identify the cause-effect aspects and the characteristic of the impact (whether cumulative or synergistic). The wide range of methods available provides methodological pluralism. The combination of suitable methods will depend on the nature of the problem, the purpose of the analysis, access to and quality of data, as well as the resources available. The combination of methods makes it possible to analyse sources, resources and effects, with a good understanding of the cause-effect relationship (SMIT and SPALING, 1999).

According to Davies (1992 *apud* PARR, 1999), a method for assessing cumulative and indirect impacts should address the following questions:

1. Set limits;
2. Evaluate the interactions between the project's environmental impacts;
3. Identify previous projects with their activities and environmental impacts;
4. Identify future projects with their potential environmental impacts;
5. Evaluate the interactions between the environmental impacts of previous projects and future projects e;
6. Determine the probability of occurrence and significance of indirect and cumulative impacts and their interrelationships.

The Multiplier Effect

The concept of the multiplier method is often introduced in EIAs. The method verifies the social benefits of implementing the project, in addition to the initial capital investment. The investment or expenditure represents income for the factors of production and this becomes the income of others, along an almost infinite chain, generating direct and indirect impacts (GILPIN, 1999).

3.3.3 Methods for Valuing Environmental Impacts

- Cost Benefit Analysis (CBA)

According to Nogueira (2000), a study aimed at applying CBA could determine its benefits using various economic valuation methods. In this way, they can be compared to their direct, indirect and opportunity costs and policies can be suggested to maximise the use of the benefits. The best project will be the one with the highest net social benefit (difference between social costs and benefits).

> "A government policy can affect the environment from the street corner to the stratosphere. Yet the costs and benefits of the environment have not always been well

integrated into the evaluation of government policy and have sometimes been overlooked entirely. Proper consideration of these effects will improve the quality of policy making" (HANLEY and SPASH, 1993).

Cost Benefit Analysis, when used exclusively for EIA purposes, has a fundamental disadvantage in that it cannot measure various environmental resources because they are intangible and therefore cannot be priced on the market, such as air quality, the value of endangered species or landscapes. This factor prevents CBA from being used as an instrument for global impact in EIA evaluation (PARR, 1999).

- Cost Effectiveness Analysis (CEA)

The inability of Cost Benefit Analysis to accommodate intangibles has led to the emergence of other techniques that claim to be able to include these resources within their calculations. The valuation of intangible resources can be achieved through Cost Effectiveness Analyses that measure consumer preferences for environmental resources (GILPIN, 1995; PARR, 1999).

- Opportunity cost

The concept of opportunity cost is also pertinent to the EIA. It is the cost of satisfying an objective, measured by the value that these resources would have had if they had been used for attractive alternatives. In environmental terms, opportunity cost is assessed in the implementation of projects and programmes, checking whether the resources earmarked for them could not have been put to better use (GILPIN, 1995).

- Contingent valuation method

Environmental goods with non-excludable characteristics, public goods or "commodities" and semi-public goods can be more easily valued, since they have some market signal (reference), either through the producer's property right or the consumer's property right. However, the basic idea behind the Contingent Valuation Method (CVM) is that people have different preferences for goods and services, and this is manifested when they go to the market and pay specific amounts for them (GARROD and WILLIS, 1999).

The major criticism, however, is its limitation in capturing environmental values that individuals don't understand, or even don't know about, since populations that aren't involved in the pollution problem tend to value its depollution less, as they aren't being affected at the time.

Contingent valuation (CV) involves calculating a value for each probable effect of each possible combination of resource uses. Then, by adding up these values for each combination, it would be possible to identify the combination that maximises social value. Carrying out these calculations is no simple task and requires a great deal of investment (BRUCE, 2006; GILPIN, 1995).

- Travei cost approach

This is another environmental valuation method that is particularly useful for assessing the economic value of natural areas or recreational areas where no price is directly stipulated. In this case, the willingness to pay for an area of contemplation is assumed to be the costs incurred by people travelling to the site (GILPIN, 1995).

The basic problems with the MCV are: a) choosing the dependent variable to "run" the regression; b) travelling for multiple purposes; c) identifying whether the individual is a resident or an occasional tourist; d) calculating distance costs; e) valuing time and; f) statistical problems (NOGUEIRA *et al,* 1998).

- Hedonic price technique

When people go to the property market to buy a property, they also consider its locational and environmental characteristics when making their choice. When they make their decision, they also take into account the perception that these characteristics arouse in them. In a way, they are "valuing" these particularities of the property (NOGUEIRA *et al,* 1998).

- Ecological evaluation

Ecological evaluations seek to identify the importance of conservation and the intrinsic value of nature, complementing and reinforcing the Cost Benefit Analysis (GILPIN, 1995).

Choosing which method to use in an EIA is a difficult task and requires a careful assessment of which environmental elements you want to classify and which interactions you want to observe. Each method has specific uses and defined formats. The way in which this method is approached and used will define the entire subsequent assessment, hence its importance within the EIA.

Chapter IV examines the methods used in the EIAs under analysis, comparing the results with two of the aspects pointed out in international literature as the most important in EIA: **the analysis of cumulative effects and the analysis of the significance of impacts,** in order to verify the effectiveness of the methods used in Brazil.

CHAPTER 4

THE EFFECTIVENESS OF AIA METHODS IN BRAZIL

4.1 The choice of evaluation parameters

In order to evaluate the EIA methods used in Brazil in terms of their effectiveness, it is necessary to demonstrate, by means of a theoretical survey, what is expected to be analysed in an EIA, comparing it with what is presented in the Brazilian EIAs studied in this dissertation.

The main focus of the environmental impact analysis stage, in which the methods serve as a didactic basis for presenting the environmental variables and their interrelationships, is to verify the relevance of the impacts of implementing and operating a project in the biotic, physical, social, economic and cultural context. The methods, in their various specificities, present the actions that will take place in the project and in each phase, the impacts and effects related to these actions, the interrelationships between each affected environment and the classification of these impacts in relation to the environmental elements, already studied in Chapter II.

Given all the information collected, the multidisciplinary technical team carries out a qualitative and subjective analysis of the impacts considered most relevant, with the aim of focusing on mitigating or compensatory measures and monitoring programmes that will mitigate the environmental changes caused by the activity, in an attempt to make the project environmentally viable.

Analysing the effectiveness of the methods presented in Brazil therefore requires an assessment of how each environmental element has been classified, how the actions, phases, impacts, effects and interrelationships are being demonstrated, comparing them with the guidelines presented in the literature on the subject. As the subject is extensive and touches on various areas of knowledge, this study concentrates on the effectiveness of the methods, under two approaches considered to be the most important: **the analysis of the significance of the impact**, as it is the most relevant step in the EIA, since it presents the conclusion of the technicians on the project and its impacts and analyses the method used under all the approaches **and the analysis of the** impacts.

cumulative impacts, the lack of which in studies is pointed out by Parr (1999) as one of the main shortcomings of Environmental Impact Assessment methods.

4.2 Definition of environmental elements

When analysing the significance and cumulative effects of impacts, one of the main steps is to classify the environmental elements, but as already described in Chapter II, there are various interpretations of what each element is, which makes it difficult to understand the subject, hindering the evaluation and effectiveness of public policy as a whole. In order to facilitate understanding of this chapter and make the concepts more precise, based on the information found in international literature, the following definitions are proposed for the environmental elements:

Table 4.1 - Definition of environmental elements

Environmental Element	Definition
Distribution of social benefit burdens	It relates to the environmental benefits or costs caused by the impacts, verifying whether the impact will cause a cost/benefit to society or whether this cost/benefit is restricted to the entrepreneur.
Magnitude	It refers to the size or degree of affectation of an impact, relating it to the intensity of its changes.
Occurrence	It is the probability of the impact occurring with the insertion of human action.
Nature	It is classified as beneficial, for impacts that promote environmental improvement, and adverse, for impacts that promote a reduction in environmental quality.
Extensionor scope	It refers to the spatial extent that the impact can reach, whether it is local or global.
Duration	It is related to the time that the impact will occur, defined in the short, medium or long term.
Reversibility	Reversibility occurs when spontaneous recovery is possible when the generating action is excluded.
Frequency	Frequency is related to the continuity of the impact, which can be permanent, temporary or seasonal. An impact with a long duration can be seasonal, for example.
Importance	The impact on biodiversity must be assessed, verifying the integrity of habitats, the state of conservation and the importance of the impacts for the socio-economy.

Influence	It refers to the primary impacts (considered direct) and the effects resulting from them (called indirect)
Cumulative properties Synergistic	Analysing the properties of impacts to add up or multiply; cumulative impacts are those that which accumulate over time and space and result from a combination of effects arising from one or more actions.
Occurrence of effects associates	The interaction of the impacts of project under study with other projects (and impacts) already taking place in the affected region.
Acceptability	This is understood as the degree of acceptance of the impacts by the affected community, taking into account public policies and local regulations. It should be carried out by means of surveys of the population and their promotion in the EIA process.
Mitigation Potential	They take into account the possible forms of mitigation, looking at the effectiveness of the measures and the investment needed to minimise the impact.

Source: Prepared by the author.

Once the environmental elements have been conceptualised, the next step in the analysis is to see how significance and cumulativeness are addressed in the Brazilian EIAs evaluated. To this end, the guidelines found in the literature on these two topics were analysed, as shown below.

4.3 Impact Significance Analysis

According to Rossouw (2003), the significance of the impact is at the heart of identifying, predicting and evaluating impacts and making decisions in EIA, acting at all stages of the process. It is understood as the result of the combination of scientific methods and the values assigned by the technical team when predicting and classifying impacts. However, the understanding of its concept is quite diffuse, being subdivided into two points of view: 1 - the significance of the impact must be analysed globally in all environmental policy processes, encompassing the preliminary and detailed analysis phases and; 2 - the significance of the impact is a stage in the development of the Environmental Impact Study, whose objective is to point out the most significant impacts so that efficient measures and monitoring can be proposed to make this impact less relevant, this being a more specific and technical view.

Focusing on the EIA, the subject of this dissertation, in a general analysis, significance can be defined as the evaluation of the impact through differentiation and demonstration (carried out by means of methods) and also the comparison of the particularities of each environmental alteration considered relevant in the prognosis. Demonstration allows the assessor to present their technical knowledge, their position in relation to the environmental diagnosis carried out and the changes caused to the environment by the activity, providing greater understanding and facilitating public participation (NEPA, 1969).

Section 1508.27 of the US National Environmental Impact Assessment (NEPA) law, established in 1969, a pioneer and the basis for other international legislation, already provided guidelines on significance and how it should be measured. According to the law, the significance of an impact must be analysed:

- The degree to which the project could affect public health or safety; the particular characteristics of the site, such as proximity to historical or cultural resources, parks, areas of agricultural importance, wetlands, rivers of scenic beauty or ecologically critical areas;
- The degree to which the effects on the quality of the human environment can be highly controversial;
- The degree to which possible effects on the human environment are highly uncertain or involve unique or unknown risks;
- The degree to which the action may set a precedent for future actions with significant effects or represents a decision in principle about a future consideration; whether the action is related to other actions whose impacts are individually insignificant but cumulatively significant;
- The degree to which the action may adversely affect districts, sites, roads, highways or objects that are listed or may be listed or may cause loss or destruction of significant scientific, cultural or historical resources, and;
- The degree to which the action may adversely affect an endangered species or its habitat and whether the action threatens to violate a federal, state or municipal law or other environmental protection requirements.

4.3.1 Assessing significant impacts: practices and guidelines

Within the context of NEPA (1969), Lawrence (2005) states that the significance analysis must be: **focused and efficient** (it must concentrate resources and efforts on essential and relevant issues), **explicit and clear** (it must present, from the stage of predicting impacts, procedures that can be easily understood), **logical and reasoned** (the entire team must be able to follow the same logic of reasoning in the analysis), **systematic and traceable** (there must be an orderly and integrated process for presenting the various characteristics presented), **appropriate** (according to the context in which it is inserted), **consistent** (equal situations should be treated in a similar way), **inclusive** (enabling the participation and understanding of all parties involved), **collective and collaborative** (where stakeholders can determine what is or isn't important), **effective** (it should demonstrate a substantive and procedural policy, achieving objectives, based on priorities) and **adaptable** (it should adapt to uncertainties and changing circumstances). A poorly carried out significance assessment has several consequences, including a lack of clarity of information and a reduction in the participation of the parties involved due to a lack of understanding of the matter.

According to Lawrence (2007) the most common faults are:

- Lack of clarity in determining value judgements;
- The inaccuracy of the information due to different interpretations and worldviews, based on the ethical and moral values of the team members and their evaluators;
- Confusion between the definition of environmental elements and their incorrect categorisation;
- Inadequacy of the project in the context of the development, as well as the failure to include the wishes of the affected population in the evaluation;
- Confusion between the significance and magnitude of the impact (according to Rossouw (2003), the magnitude of the impact is determined empirically and is just an environmental element to be observed, while significance should involve a process of determining the acceptability of the impact by society);
- It analyses the significance of an impact after mitigating and compensatory measures have been presented, which would clearly minimise its initial significance;
- It disregards the economic and social impacts;
- Insufficient consideration of the cumulative and synergistic nature of environmental impacts;
- Failures in the application of methods, with inadequate use of the different tools and procedures for reviewing and classifying impacts.

According to Lawrence (2007), Rossouw (2003) and NEPA (1969), in order to have a consistent and systematic basis for decisions on alternatives and projects, the significance analysis must be preceded by:

1. **Identification and demonstration of the actions/effects/impacts related to the various phases of the project,** as well as the interrelationships between the environmental variables (the method used must be able to demonstrate and integrate all the variables, but there is no single method, and there may be a combination of more than one method presented in chapter III) and;

2. **Presentation of environmental elements, with classification using indices that will form a value scale.** This value scale can be made up of qualitative and quantitative indices. Quantitative indices tend to be more consistent, traceable and explicit, but they make it difficult for society to participate, as they work with technical variables and quantities, making it difficult to understand the subject. Qualitative indices are more widely used precisely because they facilitate the participation of the affected community, since they make evaluation more flexible.

The minimum environmental elements that should be assessed and the scaling suggested in the literature are:

Acceptability

Table 4.2 - Description and value scale for the environmental element "acceptability"

index	Description
High (unacceptable)	Either the project has to be abandoned in part or completely, or the project can be redesigned by removing the unacceptable impacts.
Medium (negotiable)	It is only acceptable with regulatory control and the commitment of the entrepreneur, through legal sanctions.
Low (acceptable)	There is no risk to public health and it is widely accepted.

Source: Prepared by the author, adapted from Rossouw (2003).

- **Duration**

Table 4.3 - Description and value scale for the environmental element "duration"

index	Description
High (long term)	It considers the impact to be permanent and long-term, i.e. more than 15 years.
Medium (medium term)	It is a reversible impact over time, where the medium term is considered to be between 5 and 15 years.

Low (short term)	It is a rapidly reversible impact, being short term and not exceeding 5 years.

Source: Prepared by the author, adapted from Rossouw (2003).

- **Extension**

Table 4.4 - Description and value scale for the environmental element "extension"

index	Description
High	It's a wide-ranging impact that reaches far beyond the limits of the project, whether it's regional, national or global.
Medium	Affects areas close to the boundaries of the development; considered local only.
Bass	It only takes place within the development area.

Source: Prepared by the author, adapted from Rossouw (2003).

- **Intensity**

Table 4.5 - Description and value scale for the environmental element "intensity"

index	Description
High	Changes are even taking place in areas of important conservation value, or rare or threatened species are being destroyed.
Medium	The alterations occur in areas of potential conservation or natural resource utilisation, and there are also changes to the species that occur in the area.
Bass	The changes are made in areas with little conservation potential and the changes to existing species are small.

Source: Prepared by the author, adapted from Rossouw (2003).

- **Mitigation potential**

Table 4.6 - Description and value scale for the environmental element "mitigation potential"

Index	Description
High	There is a high potential for mitigating negative impacts, resulting in insignificant effects.
Medium	There is a potential for mitigating impacts, but there are still significant effects.
Bass	There is little potential for mitigation, or there are no mitigation mechanisms at all, which doesn't do much to

	minimise the impact.

Source: Prepared by the author, adapted from Rossouw (2003).

- **Occurrence**

Table 4.7 - Description and value scale for the environmental element "occurrence"

index	Description
Definitive	More than 90 per cent chance of occurrence.
Likely	Above 70% chance of occurrence.
Possible	Above 40% chance of occurrence.
Unlikely	Less than 40 per cent chance of occurrence.

Source: Prepared by the author, adapted from Rossouw (2003).

- **Significance**

The significance of the impact is also obtained through indices that reflect the team's opinion of what they have analysed. The indices suggested by the literature for significance are:

Table 4.8 - Description and value scale for "significance"

index	Description
High	It is the highest form possible within the limits of the impacts that may occur. In the case of negative impacts, no mitigation is possible that could compensate for the impact, or mitigation is difficult, expensive, time-consuming or some combination of these.
Medium	0 impact is real, but not substantial in relation to other impacts that could have an effect within the limits of those that could occur. In the case of negative impacts, mitigation is quite feasible and easily possible.
Minimum	0 impact is of low order and produces real effects of low relevance. Mitigation is easily achieved or even unnecessary.
None	Zero impact

Source: Prepared by the author, adapted from Rossouw (2003).

The technical approach to analysing significance must be effective and appropriate, emphasising technique and science, integrating the community, local public policies and regulatory standards. Care must be taken not to exclude or marginalise the public, ignore or underestimate community knowledge and interests, inhibiting dialogue and negotiation between interested and affected parties or limiting innovation and adaptation.

Professional judgements and methods (e.g. matrices, checklists, etc.) are aids in the process of deciding whether to approve a potentially polluting proposal. Methods are often dry tables that fail to fully capture relevant environmental distinctions, interactions and changes. The combination of scientific and non-scientific methods can have a collaborative approach to the process rather than a bias.

4.4 Analysing Cumulative Impacts

Cumulative impacts are defined as those effects resulting from past, present and future actions which, when combined, become significant for the environment, being represented by the sum of the impacts of the proposed project and the interactions between them (LAWRENCE, 2005; CANTER AND KAMATH, 1995).

International literature points out that analysing cumulative effects is an important factor in the EIA process, mainly because of its ability to provide information that will allow decision-makers to plan the pace of development or the total amount of development in any geographical area or region. This will be reflected in the management of natural resources, land use planning, strategic evaluation, public-private partnerships and the direction of sustainability strategies (LAWRENCE, 2005; CANTER AND KAMATH, 1995).

Although NEPA (1969) called for cumulative impacts to be included in EIAs, this assessment is still subtle and fragile. According to Canter and Kamath (1995) and Parr (1999) this is due to: 1 - the lack of methodological procedures defined for this purpose; 2 - the technical difficulties with operationalising the assessment of cumulative and indirect impacts and interactions; 3 - the lack of technical preparation by environmental agencies, which are unable to provide support for the technical team and; 4 - the greater need for time and resources.

Various actions have been taken to solve the problem. The EPA has produced manuals that discuss the assessment of cumulative impacts. International literature has also published several articles on the subject, with the main aim of directing the study of cumulative impacts through an effective method. **This effectiveness is understood as an integration of the various relevant environmental elements with the exposure of the variables, in a method capable of demonstrating the cumulative and synergistic aspects of the impacts.**

4.4.1 Assessing   cumulative impacts: practices and guidelines

Canter and Sadler (1996) and Clark (1993) describe the minimum content of a cumulative impact assessment, which should include:

1) The outline of the applicant's goals and objectives and their compatibility with the initiatives of the regulatory bodies;
2. establishing the spatial and temporal limits of the project;
3. identifying all the relevant aspects of the project;
4. the identification of other projects or activities in the project's area of influence that could cause cumulative or synergistic environmental alterations;
5. identifying the environmental factors that could be affected by the project,
6. the identification of ecosystem change thresholds, e;
7. analysing the impacts of all the alternatives, using this information to select a proposal and establishing an impact control programme.

In order to carry out this analysis, the various **environmental elements** related to the impact must be studied, as described in Chapter II. Here, in the cumulative approach, the environmental elements appear in an integrated way, where there is a combination of elements, with the aim of identifying the potentiation of the impacts that have hitherto been analysed individually.

Among the environmental elements considered essential for promoting cumulative impact analysis, Canter and Sadler (1997) and Parr (1999) point out the following:

■**Occurrence**   ;

-**Duration** , where the accumulation time will be the time interval between the disturbance and the natural recovery of the environmental system;

Scope   , addressing the geographical proximity of other projects, ...

Cumulative and synergistic properties , identifying the likelihood that they will affect the same environmental system (when the spatial proximity between disturbances is less than the distance needed to eliminate or disperse the impact);

- **Influence,** the potential of the project to have a wide influence and lead to a wide range of effects, and finally;

- **Occurrence of associated effects** due to proximity to other projects.

Once the environmental elements have been identified and studied, the next step is to organise them into a method that allows them to be analysed and integrated. As discussed in Chapter III, there are various methods in the literature, each with its own weaknesses and potential. Analysing cumulative effects according to NEPA (1969) is conceptually simple but practically difficult. Fortunately, the methods and tools available for environmental impact assessment can be used to analyse cumulative effects. However, each method has several weaknesses that must be overcome by combining other methods.

Clark (1993 *apud* Canter and Sadler, 1997) investigated traditional methods from the point of view of cumulative impacts, verifying the potential and weaknesses of each one and concluding that no method analysed encompasses all the requirements of an effective method, since it fails to interact all the aspects considered relevant to the operationalisation of cumulative impact analysis. The analysis carried out by Clark (1993) can be summarised in Table 4.9.

Table 4.9 - Description of the methods from a cumulative effects perspective

Methods Primary	Description	Potential	Weaknesses
Questionnaires, interviews and panels	Questionnaires, interviews and panels are useful for gathering a wide range of information, even identifying the multiple actions and resources needed to tackle cumulative effects. Sessions, interviews with experts and groups can help identify cumulative effects and important issues in the region.	• It's flexible; • You can work with subjective information.	• You can't quantify it; • Comparing alternatives is subjective.
Cleckists	They identify potential cumulative effects by providing a list of likely common effects or juxtaposition and multiple actions. Lists are potentially dangerous for the analyst who uses them as a shortcut to in-depth assessment and the conceptual problems of cumulative effects.	- Systematic; - Concise.	• You can be inflexible; • It does not lead to interactions or cause-effect relationships.
Matrices	Matrices are well suited to combining values in individual cells (matrix algebra thinking) to assess the actions of individual cumulative effects or multiple effects, both from ecosystems and human interactions.	• Presentation of a comprehensive comparison between the alternatives; • It covers multiple	• They don't show a time scale; • No cause-effect relationships.

		projects and their interactions.	
Network and Diagrams	Networks and diagrams are an excellent method for delineating a cause and effect relationship leading to cumulative effects. They allow the user to analyse the multiple effects of different actions and identify the indirect effects that accumulate from direct impacts.	• They direct cause-effect relationships; • Identify the indirect effects.	- It does not specify the scope of the impacts and does not target them in relation to time.
Environmental Systems Modelling	Modelling is a powerful tool for quantifying a cause and effect relationship, leading to cumulative effects. Modelling can take the form of mathematical equations describing cumulative processes or it can constitute an expert system that calculates the effect of different scenarios based on a programme of logical decisions.	• It can give unequivocal results; • They show causality; • They can be quantified; • They can integrate time and space.	• You need a lot of data; • It can be expensive.
Geographic Information System	It incorporates the locational aspect of cumulative effects, helping to define the perimeter to be analysed and the environmental systems affected. Overlapping maps can show zones where impacts accumulate.	• They focus on the spatial aspect; • It has an effective visual presentation.	• It does not explain the indirect effects; • The information is limited to the data from the charts used and does not show the magnitude of the impacts.

For Clark (1993 apud Parr, 1999) and Canter and Kamath (1995), the ideal method should encompass the following aspects:

1. Consider the **time scale and frequency of the impact,** with the time horizon being long enough to detect long-term interactions and the likely natural environmental changes over the period;

2. Explain aspects relating to the **geographical dimension of** disturbances; you should also recognise variations in spatial density because disturbances and their effects are differentiated throughout space;

3. Consider the different **types of disturbance,** i.e. disturbances that are unique or multiple, and recognise impacts that originate from multiple sources, or the same source repeated over time or space;

4. The ability to track the **accumulation process,** i.e. the processes of environmental change. It is necessary to differentiate between addition and interactive processes and incorporate a tool that aggregates the effects of each;

5. Identify, analyse and evaluate the **functional changes in an environmental system,** or a component of the system or process, following disturbance and;

6. Identifying, analysing and evaluating **structural changes in an environmental system,** or a component of the system or process, following disturbance. Structural change is essentially seen as spatial.

According to the Environmental Protection Agency - EPA (2008), most of the methods already described in Chapter III were good at describing or defining the problem, but are poor at quantifying cumulative effects. They address important aspects in considering multiple actions and multiple effects on resources, but they do not have a complete approach to analysing cumulative effects. For the authors, in order to have an effective method it is necessary to combine all the knowledge of the aforementioned methods with modern computers capable of storing, manipulating and displaying large amounts of data.

In summary, in order to assess cumulative impacts it is necessary to: **1 - Describe the actions/impacts/effects; 2 - Classify the environmental elements related to the impacts; 3 - Present the interrelationships between the impacts that will demonstrate the degree of cumulative synergism between them, linking the type of disturbance with the structural and functional alterations to the environmental systems and finally; 4 - Assess the data obtained regarding cumulative impacts and their potential and the possibility of presenting mitigation or compensation measures, with the aim of making the project environmentally viable.**

The next chapter will analyse the eight EIAs submitted to IBAMA as a prerequisite for obtaining an environmental licence, in relation to the approach to significance and cumulativeness. The analysis will consist of the following stages: 1 - Identification of the methods used (with the aim of verifying that the information considered important for analysing the two environmental elements under focus was presented); 2 - Identification of the environmental elements addressed (for this purpose, the conceptualisation used should be verified in order to identify similarities between the elements in the different EIAs); and 3 - Verification of the way in which the environmental impacts were analysed (analysing whether the data listed in the method used was interpreted correctly).

CHAPTER 5

ASSESSING ENVIRONMENTAL IMPACTS IN BRAZIL

5.1 Stages of analysis

In order to verify the existence of an analysis of significant and cumulative impacts in the Environmental Impact Studies presented in Brazil, eight projects were studied, presented to IBAMA by various Brazilian companies, all linked to the energy generation sector: one hydroelectric dam, two transmission lines, one radioactive waste deposit, one electro-energy reinforcement project, two gas pipelines and one gas treatment plant.The evaluation proposed in this chapter will be based on the actions highlighted in the literature as relevant to the evaluation. In summary, the following points will be noted:

Table 5.1 - Summary of the steps for analysing the significance of cumulative effects

Steps	Significance analysis	Cumulative effects analysis
1°	Identification of the methods used	Identification of the methods used
2°	Description and classification of environmental elements: • Acceptability; • Duration; • Scope; • Intensity; • Mitigation potential; • Occurrence.	Description and classification of environmental elements: • Occurrence; • Duration; • Scope; • Cumulative and Synergistic Properties; • Influence; • Occurrence of associated effects.
3°	Description of the impact assessment process, with the proposal of mitigating/compensatory measures and monitoring programmes.	Description of the impact assessment process, with the proposal of mitigating/compensatory measures and monitoring programmes.

Source: Prepared by the author.

The main difference between the two approaches lies in the environmental elements considered relevant. In order to verify the effectiveness of EIA in Brazil, it is first necessary to study how these environmental elements listed in Table 4.10 are, if at all, demonstrated in EIAs and what methods are used to highlight them, so that a comparison can be made to see if there is an analysis of the significance and cumulativeness of the impacts.

5.2 Identification of the methods used

One of the biggest concerns about EIA is the method used to identify and assess environmental changes. The way in which the information is presented is essential for the correct interpretation and interrelation of the information that will characterise the degree of significance of the impacts. From the point of view of cumulative effects, it is in the method used that the cause-effect relationships and interrelationships with other impacts can be identified. As for significance, it is only by exposing all the environmental elements and classifying them correctly that the analysis of significance can be accurate.

In order to assess the effectiveness of EIA in Brazil, it is therefore necessary to carry out a survey of the methods used in the studies being analysed. Table 5.2 describes these methods and also shows the sequence of disclosure used.

Table 5.2 - Identification of the methods used in EIAs in Brazil

EIA analysed	Methods used
EIA 01	Simple lists were used, with a description of the environment affected, the name of the impact, the environmental indicators and their presence in the three alternatives studied; then the interaction matrix based on the Leopold matrix was used, with a description of 39 potential impacts and 17 actions envisaged by the project. The correlation of the matrix reflects the intensity of the impact, its significance and its nature, and there is a legend in geometric figures that occupies the space of the impact/action interrelationships. There is also another matrix that lists the phases, the name of the impact and the environmental elements.
EIA 02, 03, 05, 07	In these works, only one method was used, based on the impact interaction matrix, containing the following variables: name of the impacts, phases, environmental elements, impacting action, description of mitigating and enhancing measures (for negative impacts).
EIA 04	A simple list was presented, containing the phases of the project and actions; and another list with a list of environmental indicators, specifying the environment affected and the environmental parameters altered by environmental element. This was followed by an interaction matrix

	containing the following information: identification of the impacts by environment, element and parameter altered, the phase of the project, classification of the elements, impacting action, place of occurrence and control and mitigation measures.
EIA 06	Only one interaction matrix was presented, containing: the affected environment, the related impacts, the environmental elements, the magnitude, the description of the measures, the degree of resolution and the degree of relevance.
EIA 08	A checklist was presented with the actions, media and environmental elements altered and a description of the impacts, followed by another list containing the environmental impacts and the classification of the environmental elements.

Source: Prepared by the author.

Most of the studies analysed used more than one method, with the checklist being used as the identification method and the interaction matrices as the assessment method. There is a great deal of originality and creativity in the elaboration and adaptation of methods, but without any scientific basis. Most of the time, what we see is that each team adopts its own method and uses it as a standard in its studies. As for identifying cumulativeness and significance, none of the methods used show the interrelationships between impacts, nor do they analyse significance, obtained by taking a global view of the environmental elements studied.

5.3 Environmental elements analysed and value scale adopted.

In a second approach to the analysis, in line with the second step of the scientific guidelines, a survey was carried out of which environmental elements were being observed in the studies. There was also a need to check the definition given to each term, with the aim of identifying the different terminologies used, which, as already identified in Chapter II, is due to a lack of homogeneity even in scientific literature. The list of environmental elements covered and the definitions given to them are listed in tables 5.3 and 5.4.

Table 5.3 - Environmental elements found in the EIAs

ENVIRONMENTAL ELEMENTS	EIA'S ANALYSED							
	EIA 02	EIA 02	EIA 03	EIA 04	EIA 05	EIA 06	EIA 07	EIA 08
Compatibility with government plans, projects and programmes.	-	-	-	-	-	-	-	-

Influence (direct/indirect)	-	x	x	x	x	-	-	x
Distribution of social burdens and benefits	-	-	-	-	-	-	-	-
Duration	x	-	x	x	x	x	-	x
Time scale or temporality	-	-	x	x	x	x	-	x
Scope	x	x	x	x	x	x	x	x
Frequency	-	-	-	-	-	-	-	-
Importance	-	-	x	x	x	-	x	-
Magnitude	-	x	x	x	x	x	x	-
Nature (positive/negative)	x	x	x	x	x	x	-	-
Occurrence (probability)	-	x	-	-	-	x	x	-
Cumulative and synergistic properties	-	-	-	-	-	-	-	-
Reversibility	x	x	x	x	x	-	x	x
Significance	-	-	x	x	x	-	-	x

Source: Prepared by the author.

As already pointed out in Chapter II, the most frequent environmental elements in the studies are: influence, duration, extent, temporality, magnitude and reversibility. This table shows us two extremely relevant deficiencies in relation to the elements under focus: **1 - No EIA studied addresses cumulativeness in its EIA and 2 - Significance appears in some assessments, but it is inserted as just another environmental element and not as an analysis of the relevance of impacts.**

Table 5.4 - Concepts found in the EIAs studied

Elements Environmental	Author	Meaning	Scale used
Magnitude	EIA 06	It quantifies the effects, which can be of small, medium or large magnitude. Its element is the degree of relevance, where it relates magnitude to the ease with which measures can be proposed.	- Small; - Average; - Big.
	EIA 01	Defined by the author as the intensity and significance of the impact, it refers to the degree of incidence of an impact on an environmental factor, in relation to its universe, as it is present in the area of influence, identified as: strong, medium, weak and variable intensity. The authors assign scores for each classification, from	- Strong; - Average; - Weak.

		10 to 8 for strong, 7 to 4 for medium and 3 to 1 for weak intensity.	
	EIA 03/EIA 04 /EIA 05	It refers to the degree of incidence of an impact on the environmental factor, in relation to its universe. According to the study team, magnitude is related to the size of the impact, which can be large, medium or small, depending on the intensity of the transformation of the pre-existing situation of the impacted environmental factor.	- Big; - Average; - Small.
	EIA 02/EIA 07	Magnitude refers to the degree of impact on a specific environmental parameter and in relation to that environmental factor as a whole.	- Insignificant; - Low;
		all. It can be high, medium, low or insignificant, depending on the intensity with which the environmental factor is modified. Considering that the impact may occur, it is then assessed regardless of the likelihood of its occurrence. According to the multidisciplinary team, the magnitude of a certain impact is classified exclusively by the relationship between the environmental factor in question and the activity, i.e. it does not take into account the possibility of affecting other environmental factors.	- Average; - High.
Occurrence	EIA 02/EIA 07	According to the team preparing the study, the probability or frequency of an impact will be high if its occurrence is almost certain and constant throughout the activity; medium if its occurrence is intermittent and low if it is practically unlikely to occur.	- High; - Average; - Low.
Nature	EIA 08	For the team, this element represents the influence of an action carried out on the project, resulting in a non-significant	• Significant positive; • PositiveNo

		positive change in the area. It can be categorised as: 1 - Significant positive impact (when an action carried out on the project results in a significant positive change in the area); 2 - Non-significant positive impact (when an action carried out on the project results in a non-significant positive change in the area); 3 - Significant positive impact (when an action carried out on the project results in a non-significant positive change in the area).	significant; • Negative; • Negative significant; • Undefined.
		significant negative (when an action taken in The result is a significant negative change in the area); 4 - Non-significant negative impact (when an action carried out in the undertaking results in a non-significant negative change in the area) and 5 - Undefined impact (when an action carried out results in an environmental change that is still uncertain, as it depends on the tools, methods and intensity used in the impacting action, becoming positive or negative through mitigating measures).	
	EIA 06	They only characterise them as negative and positive effects, without conceptualising them.	- Positive ; - Negative.
	EIA 01 / EIA 02 / EIA 03/ EIA 04 / EIA 05 / EIA 07	It is conceptualised as the nature of the impact. It indicates whether the impact produces beneficial/positive or adverse/negative effects on the environment.	- Beneficial; - Adverse.
Extension	EIA 08	It refers to the spatial delimitation of the impact based on the reduction between the	- Location; - Regional; - Strategic.

		causative action and the territorial extension affected. The author calls it "effect". It is classified as 1- local (when the extent of the impact reaches the surface delimited by the area of direct influence and a small peripheral portion of the land); 2 - regional (when the extent of the impact reaches the surface of the area of direct influence and a small peripheral portion of the land).	
		delimited by the functional area of influence and its catchment area) and; 3 - strategic (when the extent of the impact is on a strategic policy).	
	EIA 05	Indicates impacts whose effects are felt locally, in the immediate vicinity of the activity, or which can affect wider geographical areas, classified as regional, or even when they have a strategic characteristic, with national scope.	- Location; - Regional; - Strategic.
	EIA 02/EIA 03 /EIA 07	Indicates impacts whose effects are felt locally, in the immediate vicinity of the activity, or which can affect wider geographical areas, regional. Broad impacts on ecosystems have been classified as regional.	- Location; - Regional; . Global.
	EIA 01	Defined as the extent of the area covered by the manifestation of the effects of the impacts and classified as local, regional and global.	- Location; - Regional; . Global.
	EIA 06	The scope of the impact, whether it is localised or dispersed. Call the scope an area of incidence.	- Located; - Scattered.
Duration	EIA 08	Described as temporality, it represents the temporal form of occurrence of the environmental impact, presenting itself in a dimension that becomes gradual to the	- Temporary; - Permanent.

		different actions that produce the impacts on the environmental system.	
		generating) and 2 - permanent (when the impacting factors remain after the generating action has stopped).	
	EIA 01 / EIA 05	They only divide the impacts into permanent, temporary or cyclical, i.e. those whose effects manifest themselves indefinitely, over a set period of time or cyclically, and can occur seasonally.	- Permanent; - Temporary; - Cyclical.
	EIA 03/EIA 04 / EIA 06	It indicates how long the impact will last, and can be categorised as temporary or permanent.	- Temporary; - Permanent.
Reversibility	EIA 08	It refers to the ability of the element of the environment affected by a given action to return to its previous environmental conditions. It is classified as 1 - reversible (when after an impacting action the affected environmental object returns to the initial environmental conditions, either naturally or anthropically); 2 - irreversible (when the environmental object affected by an impacting action does not reach previous environmental conditions, despite attempts to do so).	- Reversible; - Irreversible.
	EIA 01 / EIA 02 /EIA 03/EIA 05/EIA 07	It classifies impacts according to whether their effects are irreversible or reversible. It makes it possible to identify which impacts can be completely avoided or which can only be mitigated or compensated for.	- Irreversible; - Reversible.
Time scale	EIA 08	Known as the time of occurrence, it is the accounting of the duration of the impact, once the action has been carried out that caused it.	- Immediate; - Medium occurrence

		It can be: 1 - immediate (when the neutralisation of the impact occurs after the end of the action); 2 - medium-term occurrence (when there is a need for a reasonable period of time to dissolve the impact and; 3 - long-term occurrence (when after the conclusion of the action generating the impact, it remains for a long period of time).	deadline; - Long-term occurrence.
	EIA 05	It differentiates between impacts that manifest themselves immediately after an impacting action, in the short term, and those whose effects are only felt after a period of time has passed in relation to their cause, in the short, medium or long term.	• Short term; • Medium term; • Long term.
	EIA 04/EIA 03	It differentiates impacts according to those that manifest themselves immediately after the impacting action, in the short term, and those whose effects are only felt after a period of time has passed in relation to their cause.	- Immediately; • Short term; • Long term.
	EIA 06	It indicates when the impact occurs, and can be short, medium or long term.	• Short term; • Medium term; • Short term.
Importance	EIA 02/EIA 07	Significance is associated with the degree of interference that specific actions or operational processes can have on the different environmental parameters. It takes into account not only the magnitude of the impact, but also its probability of occurrence.	• Insignificant; • Low; • Average; . High.
		A potential impact can be of potentially high magnitude with a low probability of occurrence, leading to low importance. It can thus be classified as high, medium, low	

		or insignificant, according to the degree of interference with environmental factors.	
	EIA 03/EIA 04 /EIA 05	It refers to the degree of interference of the environmental impact on different environmental factors, and is strictly related to the relevance of the environmental loss, which can be large, medium or small, to the extent that it has a greater or lesser influence on the local environmental quality as a whole.	- Big; - Average; - Small.
Influence	EIA 08	The authors call it spatialisation and it is the attribute that determines the level of relationship between the impacting action and the impact generated on the environment. It can be classified as 1 - direct (also called primary or first order impact. It results from the actions of the undertaking on the elements of the environment); 2 - indirect (it results from a secondary action in response to the previous action or when it is part of a chain of reactions, also called secondary or nth order impact).	- Direct; - Indirect.
	EIA 02/EIA 03 /EIA 04/EIA 05/EIA 07	It's how the impact manifests itself, i.e. whether it's direct (resulting from an action carried out by the project) or indirect (resulting from an accident, unexpected occurrence or a secondary impact caused by the project).	- Direct; - Indirect.
		by the main impact).	
Graude resolution	EIA 06	It refers to the degree of resolution of the measures proposed to reduce or enhance a given impact, indicating the susceptibility of the interference.	- Low; - Average; . High.

Form of interference	EIA 06	Indicates the appearance or increase of effects similar to cumulative and synergistic properties)	- Occasion; - Increase.
Significance	EIA 04	It is a combination of the levels of magnitude and importance. If the magnitude is high, the impact is very significant, if it is medium, it is significant, and if it is small, it is insignificant.	- A lot; - Medium; - Not much.
	EIA 01	It refers to the degree of interference of the environmental impact on the different environmental factors.	- Strong; - Average; - Weak.
Intensity	EIA 01	It refers to the degree of incidence of an impact on an environmental factor, in relation to its universe, as it is present in the area of influence.	• Strong; • Average; • Weak; • Variable.

Source: Prepared by the author.

Another point that demonstrates the weakness of the EIA carried out is the fact that there is a confusion of concepts and terminology, as can be seen in table 5.4. Magnitude, importance and significance are treated as synonyms, which jeopardises the significance analysis as a whole. In addition, the concepts are described in a summarised and confusing way, there is room for dubious interpretations and the value scales do not follow any scientific or technical parameters.

5.4 Description of the impact assessment process

The final step in analysing the effectiveness of the EIA is to check how the environmental impacts are assessed once they have been laid out in a method. The aim here is to check that the information has been assessed correctly, otherwise the significance of the impacts would be compromised. Furthermore, by carrying out the impact assessment procedure, it is possible with some effort to identify interactions between them, which would make it easier to analyse cumulative effects. Table 5.5 summarises the process used in the eight EIAs under analysis.

Table 4.14 - Description of the impact assessment process

EIA	Description of the evaluation process

EIA 01	1 - They demonstrate the environmental characteristics of the location alternatives by means of a list called "Integrated Analysis of Environmental Constraints" with quantitative data on some biotic and abiotic variables considered relevant by the technical team; 2 - They grouped the potential impacts and the variation of their indicators for the route alternatives in a matrix to assess the environmentally viable alternative; 3 - Once the alternative had been chosen, a dissertative description of the impacts generated by the project was carried out;
EIA 01	4 - A matrix was drawn up, similar to Leopold's, with the interventions planned by the project listed vertically and the potential environmental impacts horizontally. At this point, the environmental elements were analysed using symbology: intensity, significance and nature (positive/negative); 5 - Another matrix was drawn up, called impact valuation, where the impacts were listed vertically and the "valuation" was listed horizontally, which actually refers to the environmental elements. Within this matrix, the following environmental elements were listed: nature, reversibility, duration and scope; 6 - The distribution of impacts by medium was carried out, listing them only in terms of their nature, without presenting any other environmental element.
EIA 02 / EIA 03 / EIA 05 / EIA 07	1 - Describe the main actions that interface with the environment; 2 - Describe the potential environmental impacts, separating them by medium and by time. recommending measures; 3 - It presents a matrix, where the impacts are checked vertically
EIA 02 / EIA 03 / EIA 05 / EIA 07	1 - Describe the main actions that interface with the environment; 2 - Describe the potential environmental impacts, separating them by medium and by time. recommending measures; 3 - It presents a matrix, where the impacts considered relevant are checked vertically: nature, form, scope, reversibility and magnitude, the impacting action and the place of occurrence and the mitigating and enhancing measures.
EIA 04	1 - They list the organisation's actions using a checklist; 2 - They list the environmental indicators; 3 - They present a matrix for assessing the significance of potential impacts, correlating the environmental elements: importance and magnitude;

	4 - They describe the environmental impacts identified and relate them to the respective measures; 5 - It presents a matrix showing the impacts identified vertically and the phases of the work horizontally; environmental elements considered relevant: nature, form, scope, reversibility and magnitude, the impacting action and the place of occurrence and the mitigating and enhancing measures.
EIA 06	1 - They present a matrix called Qualitative and Quantitative Assessment, in which the impacts were assessed in relation to: nature, time of occurrence, area of incidence, duration, magnitude, degree of resolution, degree of relevance and forms of interference.
EIA 08	1 - It establishes a scale of value for adversity/significance, representing the influence of an action taken on the project, spatialisation, reversibility, time of occurrence and temporality; 2 - They draw up a matrix containing the actions related to the project vertically and the environmental elements affected, describing the environmental impacts related to the action/environmental element in the rows and columns, separated by phase;
	3 - They describe all the environmental impacts identified in the previous matrix; 4 - They present a matrix containing the impacts identified vertically and the environmental elements addressed horizontally, assigning the scale of value described above.

Source: Prepared by the author.

The way in which environmental impacts are assessed is different and does not follow any standardisation. The team defines what best suits the project under analysis and then describes a sequence of actions that are not always logical, let alone making the most relevant impacts and the measures used to minimise or compensate for them obvious. Some analysis procedures describe the impacts in a dissertative way, which seems to us to be an attempt to use the "Ad Hoc" method, but there is no identification that this is the objective.

The interactions between the impacts are not observed, they are analysed separately and the measures described are specific to these impacts. There is an attempt to include other environmental factors, such as parameters, indicators, etc., but all of this is done superficially and without presenting the theoretical basis.

5.5 The effectiveness of environmental impact assessment

It can be seen that in the EIAs discussed here there is homogeneity in terms of the methods used, always using the simple list integrated with the interaction matrix, in an attempt to add up the information provided by the two methods. There is also a standardisation of the assessment stages for the same company. The heterogeneity of presentation given to the same method (check-list and matrices) shows that each team establishes (at random) how they are going to compose the method using great creativity and then adopt this method as a standard for all their studies.

No study uses methods to identify the interplay between impacts, which would result in their cumulativeness and synergism. The phases/actions/impacts are described and assessed separately.

It is important to note that all the EIAs studied here have different names for the same environmental element (occurrence and probability, extent and scope, nature (positive/negative), influence (direct and indirect), and there are also serious conceptual errors as to the meaning of the term (significance/importance/magnitude), confusing the analysis and demonstrating little scientific knowledge on the matter.

The elements are presented and classified mechanically without any interrelationship. What's more, the attribution of the value scale seems to be done without any theoretical or scientific basis; each team chooses the scale that best suits them and often they fail to understand and operationalise it.

Some environmental elements are not even described; acceptability, for example, does not appear in any study and significance is often confused with magnitude and importance. Occurrence with associated effects is also an environmental element that is not discussed; in very few cases is there a map showing the geographical position of the project in relation to other points of interest in the region.

The only place where we can see any attempt to demonstrate the alterations and interactions between the environmental systems, as well as their structure and functionality, is in the description of the impacts, in which the team tries to explain the name of each impact, integrating it with the data collected in the environmental diagnosis. However, there is no clarity in the information and it seems that demonstrating the functioning and relationships between the environmental systems is not the aim of the description.

Cumulativeness and synergism appear on the maps presented in the environmental diagnosis, which, presented separately, show the current situation in the area. It would be possible to overlay them and thus identify the degrees of cumulativeness and synergism, but this environmental element is not highlighted anywhere in the study.

What is most evident in the assessment carried out with EIAs is that they present a huge amount of information, graphs and maps, but this information is underutilised in the assessment of environmental impacts. The overall result is an assessment carried out in a primary, mechanical way, without a scientific or technical basis, without standardisation, resulting in incipient, inaccurate and dubious information, which ends up making the EIA process ineffective, at least from the point of view of significance and cumulativeness, since it does not meet the assessment guidelines considered effective by the scientific literature and already demonstrated in Chapter IV.

The difficulties of international standardisation and homogeneity are also repeated in Brazil. The obstacles are mainly due to: the lack of standardisation of a targeted and effective method; the inexperience and lack of training of the technicians who carry out this analysis; and the lack of guidance from the environmental agency, mainly because they don't know how to carry out the analysis either.

CHAPTER 6

CONCLUSIONS

This study sought to verify how environmental impacts are assessed in Brazil from two perspectives: significance and cumulativeness. To this end, a theoretical investigation (international and national literature) and an empirical one (eight EIAs submitted to IBAMA as a prerequisite for obtaining an environmental licence) were carried out with the aim of observing what scientific guidelines exist and what was being done in practice.

What was observed was that the scientific literature itself has conceptual gaps that compromise understanding of the subject. There is no uniformity in the literature studied regarding the environmental elements to be addressed in an EIA, nor is there a uniform conceptualisation of these elements. In relation to the methods, there is no alignment in terms of classification and the authors describe the same methods with different names, and there is also confusion between what is meant by environmental impact assessment method, impact assessment technique and environmental valuation method.

The consequence of these gaps in the literature is the failures observed in practice. Several errors in the EIAs analysed are a reflection of the difficulty in obtaining accurate information from the literature.

The empirical investigation found that each team has its own way of approaching impacts, most of the time in a summarised and superficial way; there is subjectivity in the conceptualisation of environmental elements, which are used according to the understanding of each team; environmental elements are assessed randomly, without a scientific basis for assigning scales and values.

Although some EIAs attempt to combine the methods, they fail to identify indirect and cumulative impacts and the interactions between impacts; there is a lot of subjectivity in the process of classifying them, and this is considered the main weakness of using the methods, since it is through this method that their significance can be visualised and effective measures and monitoring proposed.

In the specific approach to cumulativeness and significance, it was observed that the latter, the magnitude and importance of the impacts, are treated synonymously in the EIAs. The only place where an attempt is made to assess the significance of the impact is in the description stage, where the teams explain the relationships between the impacts and the information from the environmental diagnosis in a dissertative manner. Cumulativeness and synergism only appear on thematic maps used in the environmental diagnosis, whose overlapping may, depending on the analyst's experience, demonstrate some interaction. At the actual assessment stage, there is no identification of these relationships.

In view of all the analyses carried out, it can be concluded that the Environmental Impact Assessment is ineffective in assessing the significance and cumulativeness of impacts, and that a large part of the instrument's ineffectiveness is related to: the lack of standardisation of a method, the inexperience of the technicians who draw up the EIAs, and the lack of guidance from the environmental agencies' technicians (who also lack scientific knowledge on the subject and do not have a minimum standardisation).

Establishing standardisation is the first step towards effective EIA, otherwise EIAs will continue to be a mishmash of underused information that does nothing to guarantee the management of natural resources or contribute to sustainable development. The EIA needs regulations on how to carry it out, standardising the environmental elements considered relevant and establishing minimum guidelines for the analysis. The subject is also in need of new research into the development of new methods or combinations of methods that can demonstrate the elements needed for a correct assessment and that can be used as guidelines for new EIAs.

CHAPTER 7

BIBLIOGRAPHICAL REFERENCES

BELLIA, V. **Introduction to Environmental Economics.** Brasília: IBAMA, 1996. 262 p-

BIODYNAMICS. **Environmental Impact Study of the Paulistas HPP, São Marcos River (GO/MG).** Volume 1/2. May 2005 (EIA 07)

BIODYNAMICS. **Environmental Impact Assessment of the 500 kV - Transmission Line Tucuruí/Açailândia.** February 2003 (EIA 02)

BIODYNAMICS. **Environmental Impact Study of the Caraguatatuba Gas Treatment Plant.** Volume 1/3. April 2006 (EIA 05)

BIODYNAMICS. **Environmental Impact Study of the Campinas/Rio de January.** November 2003 (EIA 03)

BISSET, R. **Methods for EIA: a Selective Survey with Casa Studies.** Paper presented at Training Course on EIA, China: 1982. n.p

BOURSCHEID S. A. **Environmental Impact Study of the Cacimbas/Catu Gas Pipeline. January 2005** (EIA 04)

BRAGA, B, et al. **Introduction to Environmental Engineering.** São Paulo: Ed Prentice Hall, 2ª Ed, 2006.

BRAZIL. **LAW 6.803 OF 02/07/1980 -** Provides basic guidelines for industrial zoning in critical pollution areas, and makes other provisions.

BRAZIL. **LAW 6.938 OF 31/08/1981 -** Provides for the National Environmental Policy, its Purposes and Mechanisms of Formulation and Application, and gives other

Measures. * Regulated by Decree 99.274 of 06/06/1990. Published in DOU 02/09/1981.

BRUCE, C. **Can Contingent Valuation solve the "Adding-up Problem" in Environmental Impact Assessment?** Environmental Impact Assessment, vol. 26. Elsevier, 2006, p. 570-585.

CANADIAN ENVIRONMENTAL ASSESSMENT AGENCY. **The significance of Social and Economic Impacts in Environmental Assessment.** 10 pp. Available at:< http://www.ceaa-acee.gc.ca/default.asp?lang=En&n=CD221BCC-1 &offset=6&toc=show> accessed on 25/02/09 at 6:00 am.

CANTER, L.W, KAMATH, J. **Impact Significance Determination - Basic Considerations and a sequenced approach.** Environment Impact Assessment Rev, n° 13, pp. 275-297, 1995. Available at: www.eiatrainning.com/CANTER- IMPACT-SIG-DETERMINATION.pdf Accessed on 25/02/2009 at 5:00 hs.

CANTER, L; SADLER, B. **A Tool Kit for Effective EIA Practice - Review of Methods and Perspectives on Their Application.** A Supplementary Report of the International Study of the Effectiveness of Environmental Assessment. International Association for Impact Assessment. June, 1997.

CHRISTA, I. **Environmental impact assessment and the pursuit of sustainable development. GEOView,** Australia. 2005. Available at: <http://www.ssn.flinders.edu.au/geog/geos/Christa.doc> Accessed: 16 November 2008.

CLARK, E.R. **Cumulative Effects Assessment: A Tool for Sustainable Development.** Council on Environmental Quality. Paper presented at the International Association for Impact Assessment Annual Conference. Washington, D. C, 1993.

CMB CONSULTORIA LTDA. **Environmental Impact Study of the Electricity Reinforcement System for Santa Catarina Island and the Santa Catarina Coast.** Volume 3, January 2006 (EIA 08)

NATIONAL ENVIRONMENTAL COUNCIL. **CONAMA RESOLUTION No. 001 of 23 January 1986.** Published in DOU 01/02/86.

CONSPLAN. **Environmental Impact Study of the 230 kV Transmission Line - Milagres/Coremas - C2.** Volume IV. 120 pp, September 2005 (EIA 01).

Davies, K. **Towards Ecosystem-based Planning: A Perspective on Cumulative Environmental Effects.** Prepared for the Royal Commission on the future of the Toronto waterfront and Environment Canada. Minister of Supply and Services, Ottawa, Ontario, 1992.

EKINS, P. **Economic Growth and Environmental Sustainability: The Prospects for Green Growth.** London: Routledge, 2000. pp.70 -114

ENVIRONMENTAL PROTECTION AGENCY (EPA). **Environmental Review Guide for Special Appropriation Grants.** EPA Publication 315-k-08-001. United States, 2008, available at http://www.epa.gov/compliance/resources/policies/nepa/environmental- review-guide-grants-pg.pdf, accessed on 08/02/2009 at 21:30 hs.

FARIA, C. A. P. **Idéias, Conhecimento e Políticas Públicas - Um Inventário sucinto das Principais Vertentes Analíticas Recentes.** Brazilian Journal of Social Sciences, Vol. 18, No. 51, 2003.

GARROD, G. and WILLIS, K.G. **Economic Valuation of the Environment: methods and case studies.** Edward Elgar: Cheltenam, UK. Nortampton, (1999).

GILPIN, Alan. **Environmental Impact Assessment (EIA).** Cambridge: Cambridge University Press, 1995. p.1-73; 169-179.

GLASSON, J; SALVADOR, N. **EIA in Brazil: a Procedures-Practice Gap. A Comparative Study with Reference to the European Union, an Especially the UK.** Environmental Impact Assessment Review, v. 20, 2000, p. 191-225.

GOMES, M.A, AMÂNCIO, R. **Políticas Públicas para o Meio Ambiente.**

UFLA/FAEPE. Lavras, 2000.

GONTIER, M; BALFORS, B; MÕRTBERG, U. **Biodiversity in Environmental Assessment: Current practice and tools for prediction.** Environmental Management and Assessment Research Group. Environmental Impact Assessment Review, vol. 26, 2006, p. 268-286.

HANLEY, N. and SPASH, C. *Cost-Benefit Analysis and the Environment.* Hants. England: Edward Elgar, 1993, 278 p.

HORBERRY, J. **Status and application of EIA for development.** Gland, Conservation for Development Centre, 1984. 86 p.

HUFSCHIMIDT, M.M; JAMES, D.E; MEISTER, A.D; BOWER, B.T;DIXON, J.A. **Environmental Natural Systems and Development: An Economic Valuation Guide.** Johns Holpins University Press. Baltimore, 1983. 338 p.

INSTITUTE OF ECOLOGY AND ENVIRONMENTAL MANAGEMENT. **Guidelines for Ecological Impact Assessment in the United Kingdon.** Pp. 53. Available at < http://www.ieem.net/ecia/impact-assess.html> accessed on 08/01/2008 at 23:27 hs

KUITUNEN, M; JALAVA, K; HIREVONEN, K. **Testing the Usability of the RIAM Method for Comparision pf EIA and SEA Results.** Environment Impact Assessment Review, vol. 28. Elsevier, 2008. p. 312-320.

LAWRENCE, DAVID. **Impact Significance Determination - Back to basics.** Environmental Impact Assessment Review, n° 27, pp. 755-769, 2007.

LAWRENCE, DAVID. **Significance Criteria and Determination in Sustainability-Based Environmental Impact Assessment.** Final Report. Mackenzie Gas Project Joint Review Panei. November, 2005. Available at: http://www.ngps.nt.ca/upload/joint%20review%20panel/specialist%20advisors/Mr%20David%20Lawrence/-51203_Lawrence_Significance_Criteria_Determination_on_Sustainability_Based_EIA.pdf accessed on 25/02/09 at 5:30 hs.

LENZEN, M; MURRAY; S, KORTE, B; DEY, C. **Environmental Impact Assessment Including Indirect Effects - A Case Study Using Input-Output Analysis.** Environmental Impact Assessment Review, vol. 23. Elsevier: 2003. p. 263-282.

LINDHJEM.H; HU, T; MA,Z; SKJELVIKJ; SONG, G; VENNEMO, H; WU, J; ZHANG.S. **Environmental economic impact assessment in China: Problems and prospects.** Environmental Impact Assessment Review, vol 27. Elsevirer, 2007. p. 1-25.

MARTIN, C; RUPERD, Y; LEGRET, M. **Urban Stormwater Drainage Management: The Development of a Multicriteria Decision and Approach.Environmental Impact Assessment,** vol. 181. Elsevier, 2007. p. 338-349.

MILARÉ, E. **Environmental impact assessment in the East, West and South:** experiences in Brazil, Russia and Germany. AB'SABER, A. N.; MÚLLER-PLATEBERG, C. (Org.) Previsão de impactos. 2. ed. São Paulo: USP, 1998. p. 53.

MONTAZ, S. **"Environmental Impact Assessment in Blangladesh: a Critical Review".** School of Applied Sciences. Centre for Sustainable Use of Coasts and Catchments. University of Newcastle. Ourimbah. NSW 2258. Australia, 2001.

MOREIRA, I.V. **Origin and Synthesis of the Main Methods of Environmental Impact Assessment (EIA).** MAIA, 1ª Edition, April 1992.

MOTA, J. A. **O Valor da Natureza: Economia e Política dos Recursos Naturais.** Rio de Janeiro: Editora Garamond, 2001.

MOURA, H; OLIVEIRA, F. **O Uso das Metodologias de Avaliação de Impactos Ambientais em Estudos Realizados no Ceará.** Available at: <http://www.ecoeco.org.br/conteudo/publicacoes/encontros/vi_en/mesa3/A_an_lise_

---

e_o_Uso_das_Metodologias_de_Avaliao_de_Impacto_Ambiental_em_Estudos_R ealizados_no_Cear_.pdf> Accessed on 15/12/2008.

MRS ENVIRONMENTAL STUDIES. **Environmental Impact Study of Unit III of the Intermediate Radioactive Waste Depot** - DIRR of the Almirante Álvaro Alberto

Nuclear Power Plant - CNAAA. Volume 04. July 2003 (EIA 06)

MUELLER, C. C. **Manual de Economia do Meio Ambiente.** Brasília: UnB/NEPAMA, 2001

NICOLAÍDES, D.C.R. **Environmental Impact Assessment: An Analysis of Effectiveness.** 2005. 84 f. Dissertation (Master's Degree in Environmental Economic Management) - Department of Economics, University of Brasília, Brasília, 2005.

NITTA Y; YODA, S. **Clallenging the Human Crisis: "The Trilemma".** Technological Forecasting and Social Change, vol: 49, 1995, pp. 175 - 194.

NOGUEIRA, J. M; PEREIRA, R. R. **Critérios e Analálises Económicas na Escolha de Políticas Ambientais.** Brasília: ECO-NEPAMA, 1999.

NOGUEIRA, J; MEDEIROS, M; ARRUDA, F. et al. **Economic Valuation of the Environment: science or empiricism.** University of Brasilia - NEPAMA, July 1998. Caderno de Pesquisas em Desenvolvimento Agrícola e Economia do Meio Ambiente.

PARDO, M. **Environmental Impact Assessment Myth or Reality? Lesson from Spain.** Environmental Impact Assessment Review, Vol. 17, n° 2. Elsiever, 1997, p. 123-142.

PARR, S. **Final Report on the Study on the Assessment of Indirect and Cumulative Impacts, as Well as Impact Interactions Within the EIA Process.** European Commission Directorate General XI, Environment, nuclear Safety and Civil Protection. Volume 1. May 1999.

PASTAKIA, C, JENSEN, A. **The Rapid Impact Assessment Matrix (RIAM) for Environmental Impact Assessment.** VKI Institute for the Water Environment. Environ Impact Assess Rev, vol. 18. New York, 1998, p. 461-482.

Principies of Environmental Impact Assessment - Best Practice. S.I.: IAIA, 1 999. Available at: http://www.iaia.org/modx/assets/files/Principles%20of%20IA web.pdf Accessed: 10 September 2008.

PUGAS, M. **Contingent Valuation of Conservation Units: Evaluating the Spontaneous and Induced DAP of the Population of Rondonópolis (MT) by the Horto Florestal.** UNB/FACE/CEEMA. Master's Degree in Economic Management of the Environment. 2006.

QUEIROZ, S.M.P. **Avaliação de impactos Ambientais:** Conceitos, Definições e Objetivos. MAIA, 1ª Edition, April, 1992.

REDEY, A and KISS, I. **Environmental Impact Assessment - A Tool For Sustainable Development.** BIOPOLITICS - THE BIO-ENVIRONMENT, VOLUME VII, A. Vlavianos-Arvanitis, L. Kapolyi (Eds.) Proceedings from the Eighth B.LO International Conference, Budapest, September 1998. Available at: < http://www.biopolitics.gr/html/pubs/budapest/redey.html> Accessed on 21 November 2008.

ROHDE, G. M. **Environmental Impact Studies.** Porto Alegre: COENTEC, 1989. pp. 42. (Technical bulletin, 4)

ROSSOUW, N. **A Review Methods and Generic Criteria for Determining Impact Significance. African Journal Environmental Assessment and Management.** Vol. 06, pp. 44-61, 2003. Available at: < http://www.ajeam-ragee.org/defaultv6.asp> accessed on 25/02/09 at 06:25 hs.

SADLER, B. **International Study of the Effectiveness of Environmental Assessment,** Final Report, Environmental Assessment in a Changing World, Evaluating Practice to Improve Performance (Canadian Environmental Assessment Agency and the International Association for Impact Assessment, Minister of Supply and Services, Canada), 1996.

SANCHEZ, L.E. **Avaliação de Impactos Ambientais: conceitos e métodos.** 1ª ed. São Paulo: Oficina de Textos, 2006.

SANKOH, O.A. Na **Evaluation of the Analysis of Ecological Risks Method in Environmental Impact Assessment.** University of Sierra Leone. Environ Impact Assess Rev. vol. 16, 1996, p. 183-188.

SANTOS, R. F. **Planejamento Ambiental: Teoria e Prática.** São Paulo: Oficina de Textos, 2004.

SANTOS, R. S; RIBEIRO, E. M; GOMES, F. G; SANTOS, L. C; RIBEIRO, M.M; SANTOS, T.C; CARIBE, D.A; SOUTO, I.M; JÚNIOR, C.D. **Compreendendo a Natureza das Políticas do Estado Capitalista.** Rio de Janeiro: RAP, 41 (5):819-34. 2007.

SPALING, H. & SMIT, B. **Cumulative Environmental Change:Conceptual Frameworks, Evaluation Approaches, and Institutional Perspectives.** Environmental Management, 17, 5, 587-600, 1999.

SPORBECK, O. **Arbeitshilfe zur praxisorientierten Einbeziehung der Wechselwirkungen in Umweltvertrâglichkeitsstudien im Strassenbauvorhaben.** Bochum, 1997.

UNITED STATES. **The National Environmental Policy Act of 1969.** As amended (Pub. L. 91-190, 42 U.S.C 4321-4347, January 1, 1970, as amened by Pub. L. 94-52, July 3, 1975, Pub. L. 94-83, August 9, 1975, and Pub. L. 97-258, § 4(b), Sept. 13, 1982.

VILA BOAS, C. **Analysing the Application of Multicriteria Decision Support Methods in Water Resources Management.** Available at: < http://www.cprm.gov.br/rehi/simposio/go/Analise%20da%20Aplicacao%20de%20Met odos%20Multicriterios%20de%20Apoio%20a%20Decisao%20na%20Gestao%20de %20Rursos%20Hidricos.pdf> Accessed on 08/08/2008.

YOUNG, C.E.F; FAUSTO, J.R.B. **Valuation of Natural Resources as a Tool for Analysing the Expansion of the Agricultural Frontier in Amazonia.** Text for discussion n° 490. Paper presented at the 1st Meeting of the Brazilian Society of Ecological Economics, Rio de Janeiro, 1997.

Zeitfracht Medien GmbH
Ferdinand-Jühlke-Straße 7
99095 Erfurt, Deutschland
produktsicherheit@kolibri360.de

Druck:
CPI Druckdienstleistungen GmbH
im Auftrag der
Zeitfracht Medien GmbH
Ein Unternehmen der Zeitfracht - Gruppe
Ferdinand-Jühlke-Str. 7
99095 Erfurt